PHYSICAL SCIENCE LABORATORY MANUAL

to Accompany Krauskopf/Beiser
The Physical Universe

PHYSICAL SCIENCE LABORATORY MANUAL

to Accompany Krauskopf/Beiser
THE PHYSICAL UNIVERSE

SIXTH EDITION

RONALD G. SAMEC

Butler University

James P. Gundlach

Clemson University

Erwin Boschmann

*Indiana University—Purdue University
at Indianapolis*

Norman Wells

Wheaton College

L. A. Youngman

*Emeritus, University of Texas,
Pan American*

McGRAW-HILL, INC.

*New York   St. Louis   San Francisco   Auckland   Bogotá   Caracas   Hamburg
Lisbon   London   Madrid   Mexico   Milan   Montreal   New Delhi
Paris   San Juan   São Paulo   Singapore   Sydney   Tokyo   Toronto*

This book was set in Press Roman Medium by J.M. Post Graphics, Corp.
The editors were Anne C. Duffy and James W. Bradley;
the production supervisor was Leroy A. Young.
The cover was designed by Charles A. Carson.
Semline, Inc., was printer and binder.

Cover Photograph
Sunset on Wyoming's Grand Tetons.
The Snake River is in the foreground.
Photograph by Michael J. Howell, *courtesy of the Picture Cube.*

Physical Science Laboratory Manual to Accompany Krauskopf/Beiser: The Physical Universe

1 2 3 4 5 6 7 8 9 0 SEM SEM 9 0 9 8 7 6 5 4 3 2 1

ISBN 0-07-035714-5

CONTENTS

PREFACE

I believe that an optimum course in the physical sciences for the nonscience major should include a rather standard laboratory component, i.e., the best introductory science courses are laboratory courses. Laboratories not only illustrate basic principles covered in the text and lectures but give the student valuable hands-on experience in "doing science." By the expression "doing science," I mean the student not only learns laboratory technique—the proper use of basic scientific instruments and the basic methods of treating and reducing experimental errors—but also experiences directly the application of several characteristics of good science. These include verifiability, workability, prediction, experimental control, and replication. Scientific models (hypotheses, theories, laws, etc.) must be verifiable—they can be operationalized and put to the experimental test. Scientific principles must also be characterized by workability—they must explain all data relating to a particular phenomena. They should successfully predict what will happen under a set of controlled conditions, and the results obtained by one experimenter should be able to be replicated by another experimenter. As a result of his or her laboratory work, the student may begin to understand, albeit in a limited way, how science "works" and gain some insight into the validity of scientific "truth." This is a very valuable commodity for a voting citizen to acquire in the wake of ever-decreasing and tightening governmental budgets and the current explosion in the amount of scientific knowledge.

I also realize that a college laboratory course may be a very difficult undertaking for a nonmathematics-oriented individual.

To aid the instructor in accomplishing the feat of teaching such a laboratory course, we have incorporated several important features into this manual. First, we have sought to keep the focus of each laboratory exercise on one particular problem. (In a few labs such as, "Calorimetry: Specific Heat and the Latent Heat of Fusion," and "The Ray Box: Reflection and Refraction," two topics are covered. In these experiments, the instructor may wish to do only one part. If this is done with the Heat Lab, we suggest that the instructor do the specific heat part and repeat the procedure with several different metal samples. In the Ray Lab, perhaps the lens and mirror section should be omitted.) Second, we have included with many of the labs a stand-alone introduction that explains the basic principle involved. Finally, we have included a worksheet-style laboratory report sheet to simplify this all-important part of the laboratory experience. In many instances, we have inserted needed equations and relationships at appropriate points in the report sheet. Basic questions follow many laboratory exercises, which test the student's understanding of the main premise of the lab.

All labs in this manual were used several times in regular college physical science courses over the last few years and have proved to be quite "doable." In particular, the physics laboratory exercises were used in a two-hour lab in which the students were required to turn in the report at the end of the laboratory period. Although the instructor does not have to follow this approach, it demonstrates that the labs included here are not too lengthy. Many students have commented that the laboratory has been the best part of the course. And students *do generally succeed* in doing the laboratory part of the course.

I would like to thank the contributors to this manual and the institutions that allowed us to teach nonscience majors and thus produce such a usable set of laboratory exercises. These contributors and respective institutions include: James P. Gundlach of Clemson University, Erwin Boschmann of Indiana University—Purdue University at Indianapolis, Norman Wells of Wheaton College, and L. A. Youngman, Emeritus, University of Texas, Pan American.

Ronald G. Samec

PHYSICAL SCIENCE
LABORATORY MANUAL

to Accompany Krauskopf/Beiser
The Physical Universe

INTRODUCTION TO THE PHYSICAL SCIENCE LABORATORY

Purpose of Laboratory Work

The purpose of the physical science laboratory is to provide hands-on experiences illustrating some basic principles of physical science—to give you the practical knowledge necessary for a well-rounded understanding of the physical sciences. In the process, you will become familiar with laboratory equipment and procedures, as well as the scientific method. This scientific approach to problems has great application outside of the physical science laboratory. You will learn techniques that will increase your confidence and ability in making meaningful future observations, in handling new and unfamiliar equipment, and in meeting new problems.

In general, the theory of a scientific principle will be presented in an experiment, and the predicted results will be tested by your experimental measurements (observations). As these principles are well-established, accepted values for certain physical quantities exist. Your purpose, basically, will be to obtain and compare experimentally measured values to accepted theoretical or measured values. The value of these experimental results will depend on your ability to make accurate measurements of physical quantities in the real world.

Procedure

I. General laboratory instructions.
   A. Laboratory equipment.
      1. A list of equipment is given with each experiment.
      2. Do not touch or turn on laboratory equipment until its use has been explained and permission has been given by your instructor.
      3. Upon completing an experiment, leave the laboratory equipment neat and in good condition for later classes.
   B. Experimental measurements and analysis.
      1. Every *measurement* you will make requires *two parts*.*
         a. A *number* (with the proper number of significant digits).

---

*Some measurements such as specific gravity (ratio of weights) and mineral hardness (relative scale) do not require units.

      b. An appropriate *unit* (data or calculated results are not physically meaningful without the appropriate unit of measurement), e.g., $l = 3.34$ cm not $l = 3.34$.

2. Record data and express results in *scientific notation* when dealing with very small or very large numbers.

      a. Scientific notation is a number between 1 and 10 multiplied by a power of 10. (Practically, place the decimal point after the first significant digit of a number and multiply by the necessary power of 10.)

      b. For example: $458{,}000$ ft $= 4.58 \times 10^5$ ft $(0.00061$ cm $= 6.1 \times 10^{-4}$ cm$)$.

3. Significant digits (figures).

      a. Most of the numbers involved in experimental work are arrived at through some process of measurement. Measurements of physical quantities do not yield mathematically *exact* numbers of unlimited accuracy and precision. Rather, the experimental measurement is approximate or uncertain by some amount. The significant digits of an experimental measurement indicate its degree of certainty. Your proper use of significant digits, then, will prevent your claiming a greater accuracy than your measurements allow.

      b. The significant digits in an experimental measurement include all the numbers (digits) that can be read directly from the instrument scale plus one doubtful or estimated number.

      c. Each of the following examples contain three significant digits:

        i. 103 g   iii. 0.0620 km

        ii. 0.00203 mm   iv. 6.74 cm

      In the 6.74-cm measurement, the 6 and the 7 were read directly and are, therefore, known with certainty. The 4, however, is estimated or doubtful since the smallest division of the scale was tenths of a centimeter (millimeters). The 4 represents a guess as to where between the 6.7- and 6.8-cm divisions the end of the measured distance lies. (*Note:* Zeros that merely serve to place the decimal point are not significant—the initial zeros in (ii) and (iii)—whereas zeros appearing to the right of figures that are already to the right of the decimal point are regarded as significant, such as the final zero in (iii).

      d. A problem case: 5430 g [the decimal point is omitted and the right-most digit(s) is (are) zero]. By the above, the number 5430 has only three significant digits, the final zero merely serving to place the decimal point, *but* the final zero *may* be physically significant in a measurement (the estimated or doubtful digit) giving four significant digits. The ambiguity is resolved by the experimenters' knowledge of the measuring instrument telling them how many significant digits are justified. Moreover, the ambiguity is resolved if the measurement is originally recorded in scientific notation as 5.430×10^3 g (or 5.430 kg), which shows explicitly the right-most zero is significant—four significant digits rather than three.

      e. Multiplication or division of two or more numerical measurements.

        i. The number of significant digits in the final answer can be no greater than that of the measurement containing the least number of significant digits.

        ii. Example: $(207.54$ ft$)(81.4$ ft$) = 16{,}900$ ft$^2 = 1.69 \times 10^4$ ft^2. This equals $16{,}893.756$ ft^2 using your calculator directly. However, you must round off your answer to the three significant digits due to the 81.4-ft. measurement.

        iii. A helpful rule to follow in such calculations is to carry one more digit during the calculations than you are justified in keeping in the final result. The example above then becomes: $(207.5$ ft$)(81.4$ ft$) = 1.69 \times 10^4$ ft^2 when rounded off to three significant digits as before. Dropping, or rounding off, digits that are not significant during computations both saves labor and prevents false and misleading conclusions.

     f. *Addition or subtraction* of numerical measurements.

       i. The final answer cannot retain any digit to the right of the first column containing an estimated digit. Consequently, all digits lying to the right of the last column in which *all* digits are significant may be dropped (rounded off).

      ii. Example:

703.5			703.5
22.0	3	should be written as →	22.0
0.0	641		0.1
+ 114.0			+ 114.0
839.5	941	should be written as →	839.6

4. Expressions for experimental error and uncertainty

     a. *Percentage error (accepted value known).* Often the object of an experiment will be to determine the value of a well-known physical quantity. The accepted or "true" value of such a quantity, A, is obtained in a standard handbook or through theoretical considerations. Through a set of experimental measurements, you will also obtain an average experimental value, E. The percentage error in the experimental value may be found as follows:

$$\% \text{ Error} = \frac{|E - A|}{A} \times 100\%$$

where $|E - A|$ means the greater value of E and A minus the lesser value—which will be called the *absolute difference* between E and A.

     b. *Percentage difference (accepted value unknown).* Sometimes it is desirable to compare the results of two experimental measurements when an accepted value is not known. The comparison, expressing the amount of agreement between results, is given as a percentage difference. This is defined as the ratio of the absolute difference—greater value minus smaller—between the experimental values, E_1 and E_2, to the average of the two results. This ratio is multiplied by 100 to express as a percent:

$$\% \text{ Difference} = \frac{|E_2 - E_1|}{(E_2 + E_1)/2} \times 100\%$$

5. Graphical display and analysis of experimental data.

     a. *The purpose of graphs.* The object of experiments is often the investigation of the manner in which one property or physical quantity depends upon another property (e.g., how the restoring force in a spring depends upon the displacement of its end). By making a plot of experimental data (measurements or observations), the relationship between two such properties is both conveniently summarized and readily seen.

     b. *Graphing procedures.* Quantities are commonly plotted using rectangular cartesian axes (x and y). Such a graph explicitly shows the manner in which the dependent variable (y, the ordinate, the "effect") on the vertical axis, depends upon the independent variable (x, the abscissa, the "cause") on the horizontal axis. The independent variable is the quantity that is undergoing controlled change by the experimenter.

  i. Label each axis with the physical quantity plotted and its units.

  ii. Choose scales for plotting such that:

      (a) The plotted data points will occupy as much as possible of the sheet of graph paper. (The same scale need not be used for both axes.)

      (b) Decimal parts of units are easily determined. This can be accomplished if each small division is made equal to one, two, five, or ten units. This avoids awkward fractions of blocks.

  iii. Choose an informative title indicating the relationship the curve is intended to show.

  iv. When two or more curves appear on the same sheet, label each by lettering parallel to the curves.

  v. Draw a smooth line (or curve) which follows the trend of the points. This means that the line need not pass exactly through each point. Indeed, "scatter," caused by various errors, will result in some data points lying on each side of the line. Your line, then, will give an average relationship of the variables—a "best fit" (Figs. 1 and 2).

c. *Straight-line graphs.* Most often, the two physical quantities (x and y) are linearly related, as is easily recognized when the points are plotted. These two quantities, then, have an algebraic relationship of the form

$$y = mx + b$$

The straight line, resulting from graphing y against x (Fig. 3) has a slope of m and a y intercept of b (the value of y at which the line crosses the y axis). The slope m equals the ratio of any set of intervals, $\Delta y/\Delta x$, from your straight line. For best results, points corresponding to data points should not be used to determine these intervals, even if they appear to lie on the line. If two quantities y and x obey the equation $y = mx$ (with $b = 0$ the straight line passes through the origin), then y is said to be *directly proportional* to x, or $y \propto x$ (Fig. 4).

d. *Other relationships reducible to straight-line graphs.* Not only is a linear relationship between two quantities x and y easily recognized, but the straight line is the only graph of a functional relationship that is really obvious. Consequently, other relationships which graph as curves, through a change of either the independent or dependent variable, are often made to be linear relations so the results can be easily analyzed.

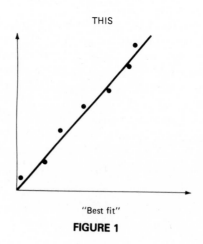

THIS

"Best fit"

FIGURE 1

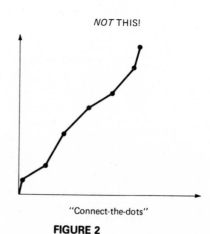

NOT THIS!

"Connect-the-dots"

FIGURE 2

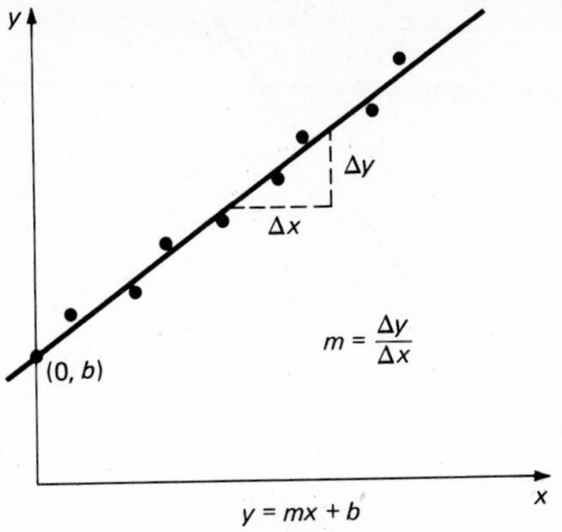

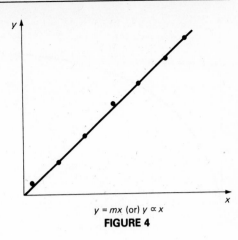

$y = mx + b$

FIGURE 3

$y = mx$ (or) $y \propto x$

FIGURE 4

i. Change of independent variable, e.g., inverse proportion (Fig. 5):

$$y = \frac{k}{x}$$

Change independent variable from x to $1/x$ (Fig. 6), or $y = k(1/x)$.

ii. Change of dependent variable, e.g., square root (Fig. 7):

$$y = a\sqrt{x} \qquad (a = \text{constant})$$

Squaring both sides:

$$y^2 = a^2x \qquad \text{or} \qquad y^2 = bx \qquad (b = a^2 = \text{constant})$$

Change dependent variable from y to y^2 (Fig. 8). The value b becomes the slope of the y^2 vs. x plot.

This "change of variables" technique will be used in several labs in this manual.

6. Trigonometric ratios.
   a. *Angle measure.* In some labs, it will be convenient to use angular measure and related ratios based on the length of sides of a right triangle. Here we will specify angular measure θ in two ways. The most common procedure is to

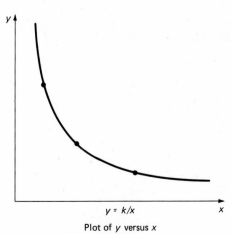

$y = k/x$

Plot of y versus x

FIGURE 5

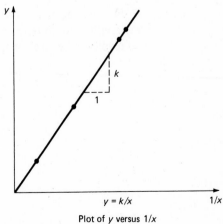

$y = k/x$

Plot of y versus $1/x$

FIGURE 6

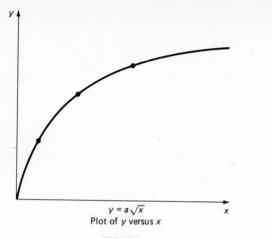

$y = a\sqrt{x}$
Plot of y versus x

FIGURE 7

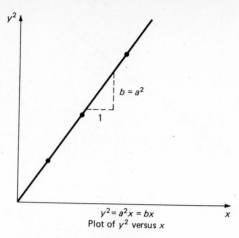

$y^2 = a^2x = bx$
Plot of y^2 versus x

FIGURE 8

divide a circle circumference into 360 equal parts, each part equal to 1 arc-degree (°). The radian (rad) is also a unit of angular measure. It is defined as the unit distance along the circumference of a unit circle. Since the entire circumference of a circle is 2π units of length ($\pi = 3.141593 \ldots$), there are 2π rad in the full 360°. Therefore 1 rad = $360°/2\pi = 57.2958°$.

b. *The right triangle*. The triangle in Fig. 9 is a right triangle since the angle at the vertex is 90°. With respect to the angle θ, the three sides of this triangle are labeled the adjacent side x, the opposite side y, and the hypotenuse r. Three basic quantities have been defined based on the ratio of the length of the sides of a triangle. They are sine (sin), cosine (cos), and tangent (tan). The definitions are

$$\sin \theta = \frac{y}{r}$$

$$\cos \theta = \frac{x}{r}$$

and

$$\tan \theta = \frac{y}{x} = \frac{\sin \theta}{\cos \theta}$$

So, we see that $x = r \cos \theta$ and $y = r \sin \theta$. Thus if the angle θ and the length r are known, both y and x can be calculated. We will use these simple results in several of the laboratories in this manual.

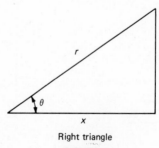

Right triangle
FIGURE 9

INTRODUCTION TO THE PHYSICAL SCIENCE LABORATORY

MATH REVIEW
REPORT SHEET

Name _____   Section _____

I. SIGNIFICANT FIGURES

1. State the number of significant figures in the following values:

   (a) 39.44 AU _____

   (b) 95° _____

   (c) 1.008 amu _____

   (d) 0.0005 in _____

   (e) 1.55×10^8 km _____

2. State the number of significant figures that the answer to each of the following problems should have:

   (a) 2.5×7 ____

   (b) $6.354/0.52$ ____

   (c) $(4.2 \times 10^8)/(2.17 \times 10^4)$ ____

   (d) $(5.85 \times 10^3) + (4.001 \times 10^5)$ ____

   (e) $(4 \times 10^7) \times (3 \times 10^6)$ ____

II. ROUNDING OFF

Round the following numbers to the indicated accuracy ("place" means significant digit):

   (a) 5653, three places _____

   (b) 0.7921, one place _____

   (c) 5.57, two places _____

   (d) 850, one place _____

III. SCIENTIFIC NOTATION

1. Put the following values in scientific notation:

   (a) 300,000 _____

   (b) 465,000,000 _____

   (c) 0.0005 _____

   (d) 0.00716 _____

   (e) 293 _____

2. Convert the following values (already in scientific notation) to normal form:

   (a) 2.19×10^4 _____

   (b) 6.8×10^3 _____

   (c) 3×10^{-6} _____

3. (a) Calculate the problems shown in I2 above. Give final answers to the proper number of significant figures. Circle your answers.

   (b) Evaluate the expression $(2.0 \times 10^6)^3$.

IV. CONVERSIONS

1. Convert 3×10^8 km/s to km/year (1 year = 365 days, 1 day = 24 h, 1 h = 60 min, etc.). (Show all work.)

2. The planet Pluto has a mean distance of 39 A.U. Convert this to km (1 A.U. = 1.55×10^8 km). (Show all work.)

3. Find the ratio of the mean distance of the moon from the earth to its size (moon's average distance = 384,400 km, moon's diameter = 3476 km). (Show all work.)

4. The earth is known to be nearly 8000 mi wide. But Jupiter is much larger, having a diameter of 144,800 km. Since 1 mi = 1.61 km, how many times larger is the planet Jupiter than the earth? (Show all work.)

V. MEASUREMENTS

What is wrong with the following *measurements?*

    (a) 0.935×10^6 m ———————————————————————

    (b) 5.75 ————————————————————————————

VI. GRAPHING

1. Take the data appearing in Table 1.1 and graph the number of sunspots (Wolf Relative Sunspot Number) for each year from 1960 to 1980. Label your graph with a title; label *both* axes with the quantity being measured *and* units measured in. Choose the best scale on each axis so that you fill your graphing area as much as possible. Draw a neat curve through the data points, making it as smooth and regular as possible. Use fine-lined rectangular graph paper, a sharp pencil, and a good eraser. *Be neat.*
2. Graph the following relationship (line):

$$y = 0.8x + 1$$

    (a) Choose any two points on the line and show that the slope is 0.8.

    (b) What is the additive constant? Why is this called the "y intercept?"

3. Repeat the above exercise with the formula:

$$y = -0.5x + 4$$

Table 1-1*

Year	Wolf Relative Sunspot Number	Year	Wolf Relative Sunspot Number
1960	112	1971	67
1961	54	1972	69
1962	38	1973	38
1963	28	1974	34
1964	10	1975	16
1965	15	1976	13
1966	47	1977	28
1967	94	1978	92
1968	106	1979	155
1969	106	1980	155
1970	104		

*Data obtained from Patrick Moore and Garry Hunt, *Atlas of the Solar System,* p. 37, and rounded to the nearest unit.

4. Draw a line of zero slope. Show that the slope is zero. What is the equation of the line you have drawn?

5. Draw a line of infinite slope. What is the equation of the line you have drawn?

6. The relationship $y = mx$ is called a *direct proportion*. For $m = 2$, graph the relationship. Show that this is indeed a direct proportion (what happens when x is doubled, tripled, halved, etc.?).

7. From the equation graphed in item 6, determine the angle θ of the generated triangle by using a protractor (use $x = 4$ as the right side of the triangle). Find $\sin \theta$, $\cos \theta$, and $\tan \theta$ with your calculator. Check your results from your calculator by making direct measurements of the sides of the triangle with a centimeter ruler.

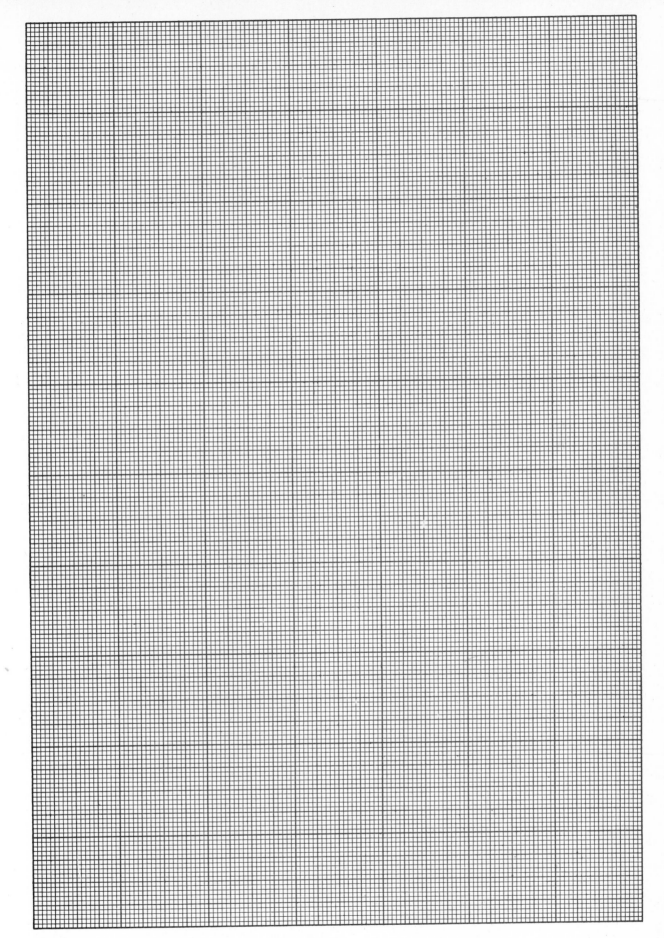

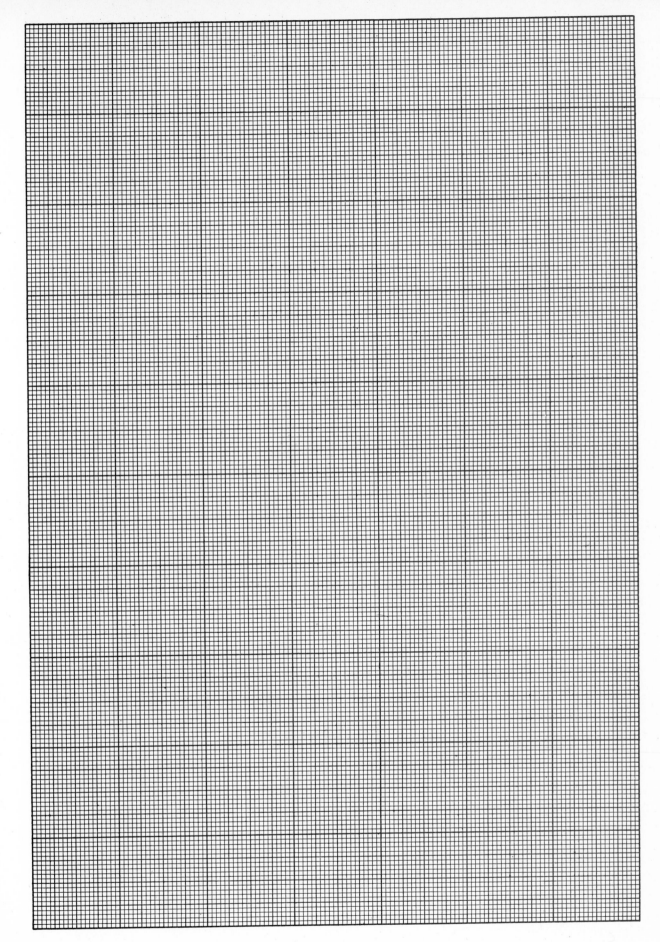

13

EXPERIMENT 1

MEASUREMENT: LENGTH, MASS, VOLUME, AND DENSITY

Materials Needed

Meter stick
Vernier caliper
Laboratory balance
Graduated cup
Blocks of wood
Irregular shaped objects
Solid balls made of various materials

Background

The physical sciences, being both quantitative and experimental, are based upon precise measurement. This is most true in the physical science area called *physics*. A scientific measurement consists of a number and a unit. The three fundamental units of mechanical measurement are length, mass, and time. Many other units are *derived* units since they are written as some combination of the fundamental units. The volume of an object, for example, is derived from the fundamental unit of length—i.e., from the dimensions of the object. Another important derived quantity and useful material property is the density of the object, which gives the relationship of its mass to its size (volume).

In this experiment, the dimensions of various regularly shaped objects will be measured by means of a meter stick and also by means of a more precise and accurate measuring instrument, the vernier caliper. From these measurements, the volumes of objects will be calculated. The volume of an irregularly shaped object will be determined by the displacement method. The mass of each object will be found with a pan balance, and the density of the material of which each object is composed will be calculated.

The smallest subdivision marked on the meter stick, or on any instrument scale, is called its *least count*—the smallest (least) unit that can be read without estimating. The least count of a meter stick is usually millimeters (mm). For precise measurements, the scale must be read to a fraction of its smallest subdivision—i.e., one digit is estimated. A measurement, therefore, may have one more significant digit than the least count of the instrument scale.

For a meter stick, with a least count of 1 mm, a measurement reading can be made (estimated) to 0.1 mm.

For more accurate and precise length measurements, a vernier caliper is used. The vernier caliper consists of a fixed rule with a main graduated scale and a movable jaw with a graduated vernier scale which slides along the main scale. The vernier scale helps in reading accurately the fractional part of the smallest scale subdivision. The least count of a vernier is 0.1 mm. Your lab instructor will further explain the use of the vernier caliper.

The mass M of a substance can be measured using a laboratory (chemical) balance—a pan balance for our experiment. The pan balance balances the weight of an unknown object against the weight of known masses which are placed in the right pan. The common pan balance has a least count of 0.01 gram (g) mass. Thus, if great care is taken, a measurement can be made (estimated) to 0.001 g.

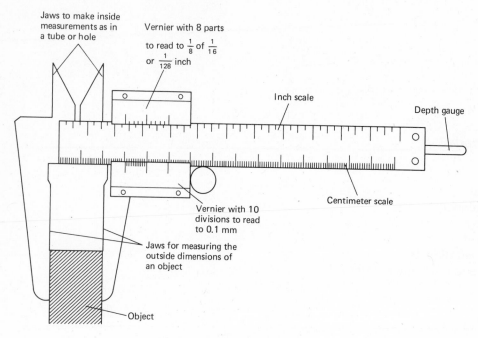

VERNIER CALIPER

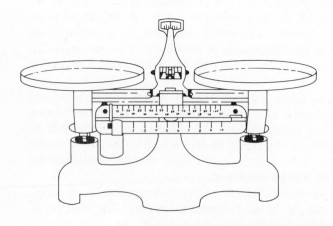

LABORATORY BALANCE

The volume V of a regularly shaped object can be calculated from the measurements of the characteristic dimensions of the object. The following formulas for volume will be needed in this experiment:

For a rectangular solid:

$$V = lwh \tag{1}$$

where l = length, w = width, and h = height.

For a sphere:

$$V = \tfrac{4}{3}\pi r^3 \tag{2}$$

where r = radius.

For a circular cylinder:

$$V = \pi r^2 h \tag{3}$$

where r = radius of circular base and h — height.

The volume V of an irregularly shaped object, such as a piece(s) of rock, is determined by immersing it in a liquid (water) in a graduated vessel. Since the object will displace a volume of water equal to its own volume, the difference in the level readings before and after immersion gives the volume of the object. Graduated cylinders typically have scale divisions of 5 or 10 milliliters (mL). If great care is taken, a measurement can be made (estimated) to 1 mL. (*Note:* 1 mL = 1 cm^3 = 1 cubic centimeter.)

The density ρ of a substance is defined as its mass per unit volume:

$$\rho = \frac{M}{V} \tag{4}$$

Density is commonly expressed in g/cm^3 but also in kg/m^3 in SI units (1 g/cm^3 = 10^3 kg/m^3). Density is a characteristic property of a substance and can often be used to identify unknown materials. Also, if the density and volume of an object are known, using Eq. (4) its mass may be obtained.

Procedure and Analysis of Data

Enter all data and calculated results in the table provided in the Report Sheet.

I. REGULARLY SHAPED OBJECTS

1.   A block of wood
     (a)   Using your meter stick, measure the dimensions of a small block of wood and calculate its volume. (All data and calculations must be written with the proper number of significant digits.)
     (b)   Using a laboratory balance, obtain the mass of the block of wood.
     (c)   Calculate the density of the block using Eq. (4).
     (d)   Determine the mass of a cylindrical log, made of the same material as your block of wood (identical physical properties), which is 10.0 m long and 80.0 cm in diameter.

2.   A marble
     (a)   Using your vernier caliper, measure the diameter of a marble and calculate its volume.
     (b)   Obtain the mass of the marble.
     (c)   Calculate the density of the marble.
     (d)   Using the table of densities given in your text (and common sense), identify the material of which the marble is made and record on your Report Sheet.

II. IRREGULARLY SHAPED OBJECT(S): PIECES OF ROCK

1.   Using a graduated cup and water, determine the volume of four or five pieces of rock by the displacement method previously described.
2.   Obtain the mass of these pieces of granite.
3.   Calculate the density of granite.
4.   Determine the mass of a tombstone made of the same material as these pieces of rock and having dimensions of 260 cm by 90 cm by 40 cm.

EXPERIMENT 1

MEASUREMENT: LENGTH, MASS, VOLUME, AND DENSITY
REPORT SHEET

Name _____   Section _____

Material	Object	Length, cm	Width, cm	Height, cm	Diameter, cm	Mass M, g	Volume V, cm^3	Density ρ, g/cm^3
Wood	Block of wood				XXX			
	Log		XXX	XXX				
____ (Fill in)	A marble	XXX	XXX	XXX				
Granite	Pieces of rock	XXX	XXX	XXX	XXX			
	Tomb-stone				XXX			

EXPERIMENT 2

MOTION: VELOCITY AND ACCELERATION

Materials Needed

Acceleration timer (buzzer unit)
Carbon paper disks
Paper tape
Metal washers
Paper clips
Masking tape
Meter stick

Background

Motion, the change of position, characterizes everything in the physical world and hence is of great interest to the understanding of the natural world. The present experiment will familiarize you with the physical quantities used to deal with the many kinds of motion. In physics the study of motion without regard to its cause is called *kinematics*.

Velocity is the time rate of change of position (displacement) in a particular direction. Having both magnitude (the speed) and direction, velocity is a vector quantity. Since this and the next experiment will study motion in only one direction (linear motion), the vector nature of velocity will not be considered. The average velocity $\bar{v}$ of an object traveling a distance Δd in a time interval Δt is defined as

$$\bar{v} = \frac{\Delta d}{\Delta t} = \frac{d_2 - d_1}{t_2 - t_1} \tag{1}$$

or in terms of distance divided by time. Measurements of average velocity are readily obtained by measuring the distances an object moves in successive equal time intervals.

Since successive distances are often not constant, velocity may change also. In this case, the body is said to be *accelerated*. Just as velocity is the rate of change of displacement with time, acceleration is the rate of change of velocity with time. The average acceleration $\bar{a}$ of an object undergoing a change of velocity, Δv (whether an increase or a decrease), in a time interval is

$$\overline{a} = \frac{\Delta v}{\Delta t} = \frac{v_2 - v_1}{t_2 - t_1} \qquad (2)$$

or distance divided by time squared. Again, the vector nature of acceleration is not considered here. Measurements of average acceleration are readily obtained by measuring the changes in velocity an object undergoes in equal time intervals.

If in each succeeding time interval, no matter how small, the object covers the same distance as before, it is moving at a constant speed and has zero acceleration. If the speed varies (increases or decreases) with time in a uniform way, the object is moving with constant acceleration. For such uniformly accelerated motion, the change in velocity of the object in each successive time interval, no matter how small, must be constant. A plot of the velocity versus time, in this case, would thus yield a straight line with a constant slope equal to the value of the acceleration.

This experiment will measure the average velocities and average acceleration of a falling object undergoing uniformly accelerated motion. Since the object is falling under the influence of gravity only (neglecting air resistance and friction), the value of the acceleration obtained will be the acceleration due to gravity, g. (Do not confuse with g, used for gram.)

General Procedure. An object (a plummet) is released from rest and allowed to fall freely for a distance of several feet. Its position during the fall is recorded at successive equal intervals of time by means of a dot record on a paper tape attached to the plummet. The dots on the tape represent a record of position and time of the falling plummet, and enable the velocity and acceleration to be calculated.

The position and time record is placed on the falling paper tape by a acceleration timer which buzzes at a constant repetitive rate. A screw-striker attached to the buzzer strikes down onto a carbon-paper disk, positioned above a paper tape, at this same constant rate. As the carbon paper disk is struck, it leaves a dot record on the moving paper tape. The space between consecutive dots represents a constant time interval regardless of the distance of separation between the dots. The distance between the dots is Δd, the distance the tape moved during the constant time interval. This distance Δd, divided by the time interval Δt, is the average velocity $\overline{v}$ during that time interval (see Eq. 1). To reduce the time required for the experiment and to minimize error, the time interval Δt will be taken as two dot-repetition intervals instead of one. This will be called *one two-tick* (1 tt). At the end of the experiment, two-ticks will be converted to seconds.

Procedure

1. Attach several (two to four) metal washers (the plummet) to the end of a paper tape, of approximately 1–1½-m length, using a paper clip and masking tape. Thread the paper tape through the slots in the buzzer unit and temporarily hold the plummet at drop height as shown in Fig. 1.

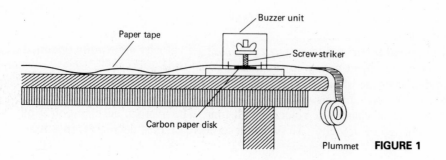

FIGURE 1

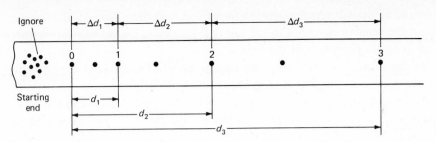

FIGURE 2

2. Start the buzzer unit and release the plummet, causing the paper tape to be pulled past the screw-striker and to leave the position—indicating dot record.
3. Stretch the completed paper tape flat upon the table top and anchor it down at each end with masking tape.
4. Ignoring the group of dots at the starting end of the tape, which were made before the plummet fell, select one of the first clear dots as the initial point for your analysis, and lable it zero as in Fig. 2. Now mark and label each second dot 1, 2, 3, etc., as shown, until the last possible complete interval has been established on the tape. Ignore the group of dots at the end of the tape; they will have occurred after the plummet hit the floor.
5. Since the two-tick is our unit of time, the time interval numbers 1, 2, 3, etc., are also the total elapsed time t in two-ticks from the zero mark to the end of that interval and are entered in column 1 of your Table 1 constructed after the example given below.
6. Measure the total distances fallen d_1, d_2, d_3, etc., between the zero mark and succeeding numbered dots and enter in column 2 of Table 1.

Analysis of Data

1. Subtract each of the total distances fallen d_1, d_2, d_3, etc., of column 2 from the one immediately following it and enter in column 3. This difference gives the distances Δd_1, Δd_2, Δd_3, etc., fallen during successive equal time intervals.
2. Divide these distances Δd in column 3 by the time interval, Δt (1 tt), to obtain the average velocity $\bar{v}$ for each interval and enter the result in column 4.

Table 1

1 Elapsed Time, t, tt	2 Total Distance Fallen, d, cm	3 Distance Fallen in One Time Interval, Δd, cm	4 Average Velocity during One Time Interval, $\bar{v}$, cm/tt
1	$d_1 =$	$\Delta d_1 =$	$v_1 =$
2	$d_2 =$	$\Delta d_2 =$	$v_2 =$
3	$d_3 =$	$\Delta d_3 =$	$v_3 =$
$\vdots$	$\cdots$	$\cdots$	$\cdots$

3. Make a plot of average velocity $\bar{v}$ versus elapsed time t from your data. *Note:* If the velocity values either remain the same or decrease in the last few intervals, ignore them since these will have occurred after the plummet hit the floor.

4. Obtain the slope of the graph and enter on your Report Sheet. This is your experimental value a_E for the acceleration of the plummet in cm/tt^2.

5. Ignoring air resistance and friction, this acceleration value also corresponds to the acceleration due to gravity. Convert the value of a_E in cm/tt^2 and enter on your Report Sheet. The buzzer unit makes 60 dots a second (60 cycles/s alternating current) and, therefore, *one* time interval equals $\frac{1}{60}$ s. One two-tick interval, then, equals $2 \times \frac{1}{60}$ or $\frac{1}{30}$ s (conversion factor: 1 tt $= \frac{1}{30}$ s).

6. Calculate the percent error of your experimental value of acceleration and enter on your Report Sheet.

EXPERIMENT 2

MOTION, VELOCITY, AND ACCELERATION
REPORT SHEET

Name _____   Section _____

Problems

(Show all necessary calculations below.)

1. $a_E =$ _____ cm/(tt)^2 (slope of graph)

2. $a_E =$ _____ cm/s^2

3. Percent error of $a_E = \left| \left| \dfrac{a_E - g}{g} \right| \right| \times 100 =$ _____

4. Why does the plot of v versus t not pass through the origin?

5. What *measurable* data are recorded on tape?

   (a)

   (b)

6. If you were to graph acceleration versus time for your plummet, what would be the shape of the curve? (Make a small sketch.)

7. Would giving your plummet an initial downward velocity different from zero result in a different observed value of the acceleration? Why?

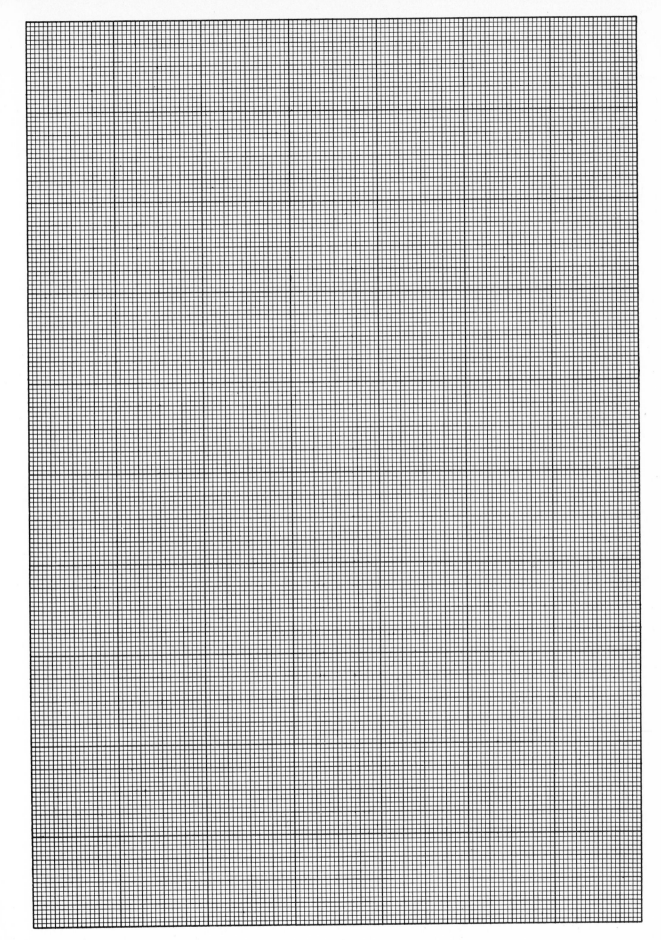

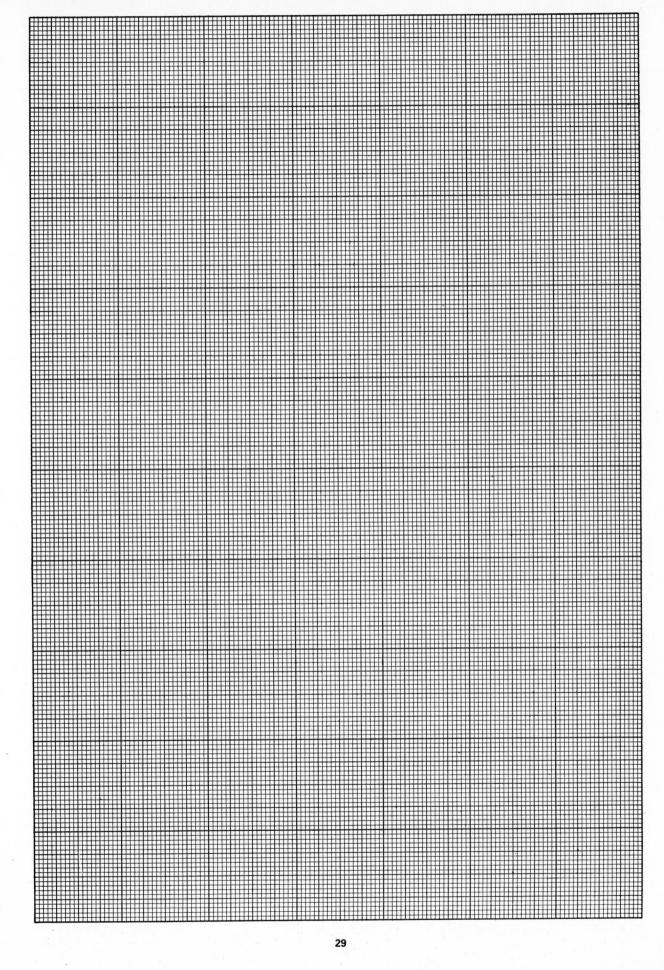

EXPERIMENT 3

THE AIR TRACK

Materials Needed

Air track
Cart
Spark timer
Tape
Meter stick

Background

Galileo attacked the problem of the motion of falling bodies by "diluting" the effect of gravity through the use of an inclined plane. We will seek to duplicate his results using a modern piece of equipment called an *air track*. The air track is used to reduce friction to near zero so that the measured acceleration is due only to the effects of gravity acting on the inclined plane. A spark timer will be used to produce dots on a strip of paper at equal time intervals. Galileo concluded that $d \propto t^2$; thus $d = kt^2$, where d is the distance traveled and k is a constant. Thus the body undergoes a constant acceleration.

We know when an object is moving with a constant velocity its position is described by

$$d = v_0 t$$

or

Position = (velocity) (time)

where d is the total distance traveled. The velocity v_0 is the constant of proportionality in the relation $d \propto t$, and the position d is constantly changing with time t. This is true when no acceleration is being applied.

When an object is moving with a constant acceleration from rest ($v_0 = 0$), its velocity is described by

$$v = at$$

Its distance of travel is

$$d = \tfrac{1}{2}vt$$

where $\tfrac{1}{2}v$ is the average speed during time t. By combining these formulas we obtain

$$d = \tfrac{1}{2}at^2$$

(Position) $= \tfrac{1}{2} \times$ (acceleration) $\times$ (time squared)

where d is the total distance traveled in time t from the point of release. Thus we see that d is proportional to t^2, where the constant of proportionality is $a/2$.

The action of gravity on an object *sliding* down a frictionless plane is shown below.

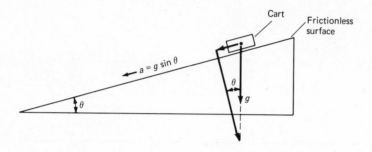

We see that the actual acceleration experienced by the cart is $g \sin \theta$ where θ is the elevation angle.

Procedure and Analysis of Data

1. Using the spark timer and special tape, launch the cart across the level track. A set of dots will be generated by the sparks as they ground to the air track. Remove the tape, and stretch it flat upon the table top and anchor it down at each end with masking tape. Mark the first clear position as zero and mark 10 more well spaced positions as 1, 2, 3, 4, . . . , 10. Place the meter stick so that it is aligned with the dots and measure the positions of each point. Subtract the initial zero position from each and record the resulting lengths 1, 2, 3, . . . , 10 in Table 1 on your Report Sheet.

2. Count the number of time spaces (ask your instructor what the frequency f is) and record this in Table 1 of your Report Sheet. Multiply each count by $1/f$ and record. This is the actual time for the respective recorded positions.

3. After completing Table 1, make a plot of distance versus time. Fit a line to the points and find a slope. This slope represents the velocity of the cart. If the points fit the line well, we can say that the air track was level and that no component of acceleration was acting on the cart.

4. Next, elevate the track and release the cart from rest. Be careful to mark the point of release on the tape. Remove the tape and measure it as before. Use the point of release as the zero position. Record all data in Table 2 of your Report Sheet.

5. Plot d versus $\tfrac{1}{2}t^2$; since $d = a(\tfrac{1}{2}t^2)$, the slope of the plot of these values will represent the acceleration.

EXPERIMENT 3

THE AIR TRACK
REPORT SHEET

Name _____   Section _____

Table 1   Initial position $x_0 =$ _____ cm; frequency $f =$ _____ s

	Position x, cm	d, cm	No. of Time Spaces	× 1/f = seconds
1.				
2.				
3.				
4.				
5.				
6.				
7.				
8.				
9.				
10.				

Table 2   Release point $x_0 =$ _____ cm

	Position x, cm	d	No. of Time Spaces	t, s	$\frac{1}{2}t^2$, s^2
1.					
2.					
3.					
4.					
5.					
6.					
7.					
8.					
9.					
10.					

1. Velocity obtained from the first plot v_0 = _____.

2. How can we be assured that the air track was level in this determination?

3. The air track is far superior to using techniques used by Galileo (rolling objects down an inclined plane). Why?

4. Acceleration obtained from second plot a = _____.

5. Was the distance traveled proportional to the time squared? Write the equation the describes the motion of the cart. What is the constant of proportionality?

Optional Questions

6. Determine the value of g (g has a known value of 981 cm/s^2).

   g = _____   PCT error _____

7. What would have happened if we released the cart on an elevated track with an initial forward velocity v_0? Write the equation of motion that gives the distance traveled.

8. A car accelerates uniformly at 13.5 ft/s^2, from a beginning speed of 50 ft/s. How far does it travel in 5.8 s? Show your work.

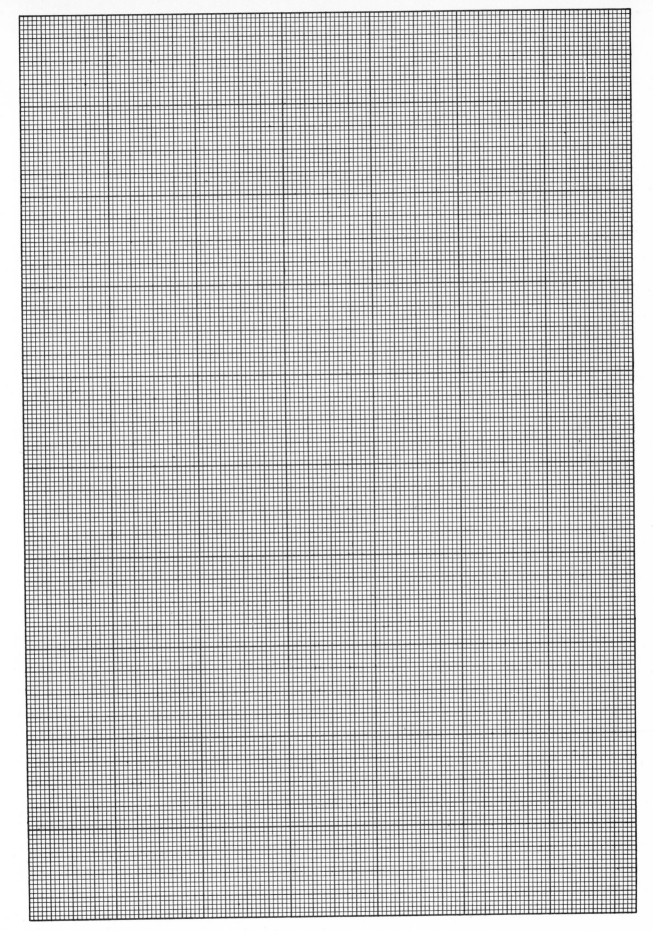

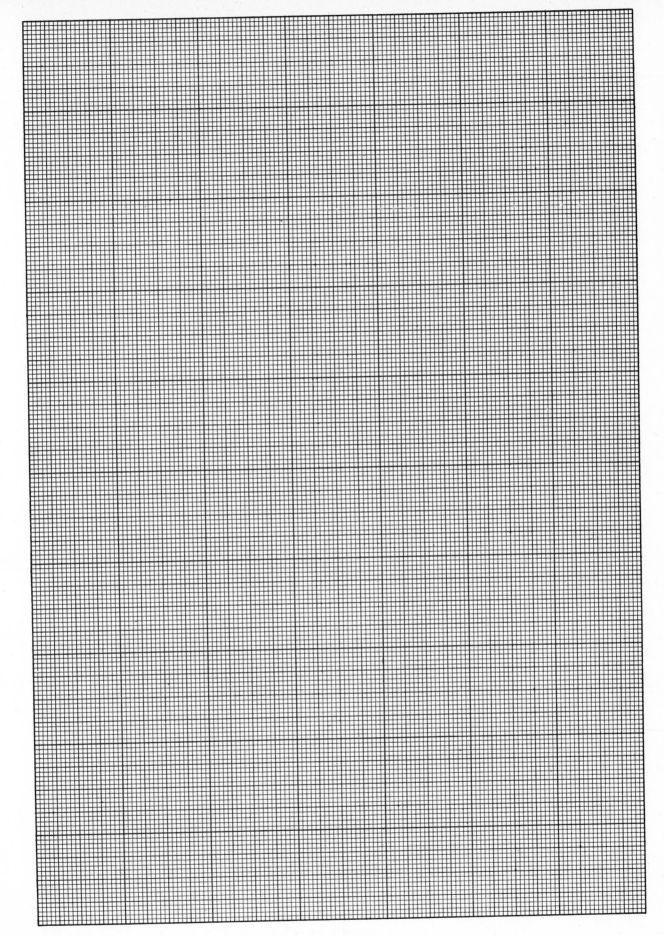

EXPERIMENT 4

UNIFORM CIRCULAR MOTION

Materials Needed

Sargent Welch centripetal force apparatus (mechanical rotator)
Stopwatch
Meter stick
Slotted masses and weight hanger
Laboratory balance

Background

Circular motion at constant *speed* is a special case of accelerated motion. It is difficult, at first thought, to see how something can be accelerated and yet have *constant* speed. However, according to Newton's first law of motion an object having no external force exerted on it (and hence no acceleration) will travel in a straight line. Now a circular path is certainly far from straight so there must be a force and a resulting acceleration in order for an object to travel in a circular path. Remember, also, that acceleration was defined in terms of a change in *velocity* and while the magnitude (speed) may remain a constant, the direction is most certainly changing as the object travels about its circular path. Figure 1 shows the velocity vector of an object traveling in a circular path at constant speed at eight successive points around the circle. The object's acceleration is, by definition, $a = \Delta v / \Delta t$. Let us take a closer look at Δv between points 1 and 2. The velocity vector v_1 plus the change Δv should equal the vector v_2. Figure 2 shows this arrangement. Note that Δv points approximately toward the center of the circle. Indeed, if we have chosen our points 1 and 2 very close together, the change in velocity Δv would have pointed directly at the center of the circle. This points out another special aspect of circular motion at constant speed, namely, that the force producing the required acceleration *must* be directed toward the center of the circle. This center directed force is called the *centripetal force*, centripetal being Latin for "moving toward the center."

A relationship between the centripetal acceleration and the object's speed and the radius of the circle can be obtained by using plane geometry. This relationship is

$$a = \frac{v^2}{r}$$

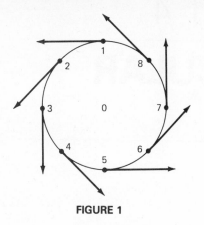

FIGURE 1

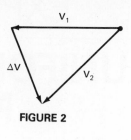

FIGURE 2

Now Newton's second law ($F = ma$) requires that the centripetal force be $F_c = m\,v^2/r$. It should be noted that a does not depend on the mass of the object, but the centripetal force does. The more massive the object, the larger the force required. Say that a time T is required to make N revolutions. How far has the rotating mass traveled in this time T? It has gone around the circle N times. To go around once, it travels a distance of $2\pi r$ (the circumference) where r is the radius and π is the constant 3.1416. Thus the total distance traveled is $2\pi r \times N$. Now the speed is the distance traveled divided by the time, or

$$v = \frac{2\pi rN}{T}$$

As we have seen, $F_e = mv^2/r$; hence $F_e = (m/r)(2\pi rN/T)^2$, or $F_e = 4\pi^2 mN^2 r/T^2$.

A good example of circular motion at constant speed can be found in many of the artificial satellites (as well as many of the natural ones) which we place in orbit about the earth. We often hear that the astronauts in orbit are "weightless" and from this "fact" it is further inferred that since they are weightless, gravity does not act on them. But remember, if no force acts on them they move in a straight line, not in a circle. The astronaut is really falling around the earth with gravity providing the centripetal force.

Procedure

A mechanical rotator will provide a mass moving in a circle at an adjustable speed (Fig. 3). The mass is connected, via a spring, to the center of rotation. As the speed in increased, the spring is stretched until the proper force is provided. The magnitude of this force may be obtained by measuring the force required to stretch the spring (when at rest) the same amount as it was stretched when in motion. Complete steps 1 through 8 for each trial separately.

1. In any given trial the position of the crossarm and radial indicator must be such that the heavy mass hangs freely exactly over the indicator when the spring is detached. Therefore, when changing the radius of rotation, both the indicator and the crossarm must be moved correspondingly. The location of the counter balance on the crossarm is not critical. Disconnect the spring and adjust the crossarm so that the hanging bob is directly above the pointer, and measure r from the center of the post to the pointer. Record this value under trial 1 on your Report Sheet.
2. Rotate the system by applying torque with the fingers on the knurled portion of the shaft. With a little practice the rate of rotation can be adjusted to keep the mass passing directly over the indicator (Fig. 4). Adjust the speed so that the tip of the bob passes right over the

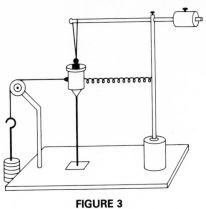

FIGURE 3

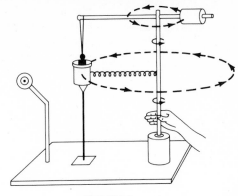

FIGURE 4

pointer each time. Measure the time T for 50 rotations. Repeat once and enter the average on the Report Sheet (trial 1).

3. Calculate and record the speed of the bob from $v = (2\pi r \times 50)/T$.
4. Measure the mass of the bob.
5. Calculate the centripetal force on the bob in rotating using $F_c = mv^2/r$.
6. Connect a string to the bob on the opposite site from the spring, run it over the pulley, and attach a weight hanger. Place enough mass on the hanger so that the spring is stretched the same amount as when the mass was rotating and so that the bob tip is again right over the pointer. Record the mass total (hanging mass) on the Report Sheet.
7. Find the weight of the hanging mass using $W = M_h g$.
8. Compare the centripetal force from step 5 with the weight of the hanging mass of step 7. Repeat steps 1 through 8, but for trial 2 add 50 g to the bob and leave r the same, and for trial 3 change r and leave the bob mass the same.
9. Why should the forces in steps 5 and 7 be equal?

EXPERIMENT 4

UNIFORM CIRCULAR MOTION
REPORT SHEET

Name _____   Section _____

	Trial 1	Trial 2	Trial 3
1. r, cm radius of motion			
2. T, s (50 revolutions)			
3. v, cm/s speed of bob $\dfrac{2\pi r \times 50}{T}$			
4. m, g mass of bob			
5. F_c, dynes centripetal force (mv^2/r)			
6. M_h hanging mass			
7. W weight of hanging mass ($W = M_h\, g$)			
8. Percent difference $[(F_c - W)/\text{average}$ $\times\ 100]$			
Answer to question 9			

EXPERIMENT 5

SIMPLE HARMONIC MOTION

Materials Needed

Spiral spring and rigid support
Weight hanger and weights
Meter stick
Laboratory timer or stopwatch

Background

Many events in nature are periodic, with a certain motion repeating itself over and over again after equal intervals of time. Simple harmonic motion is one of the most common types of periodic motion. In order to vibrate with simple harmonic motion, an object must be acted upon by a somewhat specialized variable force called *a linear restoring force*. Such a force arises in an *elastic* body which resists being deformed or distorted by a force and which returns to its original shape and size after being deformed. The object of this experiment is to study the elastic behavior of a spring and also to investigate the simple harmonic motion which results when a mass hung from the end of the spring is displaced from its equilibrium position and released.

Waves are another type of periodic motion in which a periodic disturbance spreads out from a source carrying energy with it. Mechanical waves are waves that travel through matter. Not surprisingly, periodically vibrating systems undergoing simple harmonic motion are capable of generating such waves. Examples include vibrating strings, sound waves traveling the air, etc. The simplest illustration of simple harmonic motion is the oscillatory movement of a mass attached to the end of a spring. A study of this simple system is preparatory to the understanding of wave motion.

According to *Hooke's law*, when an elastic body, such as a spiral spring, is subjected to a force which either elongates or compresses it, the displacement x from the equilibrium position is directly proportional to the magnitude of the stretching force F. Expressed mathematically

$$F = -kx$$

where the constant of proportionality k, the force per unit elongation (N/m), is called the *spring*

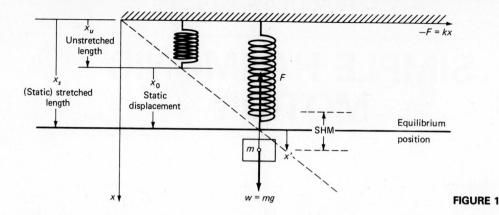

FIGURE 1

constant and is a measure of the stiffness of the particular spring—the greater the k, the greater the stiffness. The minus sign indicates that the force and displacement are in opposite directions—the force acting to "restore" the body to equilibrium. Hooke's law, therefore describes a *linear restoring force*.

As illustrated in Fig. 1, if an initially unstretched spring having a length x_u and a suspended mass m stretches the spring so that its length is x_s, then in static equilibrium the weight force w is balanced by the spring force F and

$$F = -k(x_s - x_u) = -mg \tag{1}$$

Or, letting $x_0 = x_s - x_u$ be the static displacement of the spring, Eq. (2) can be written as

$$kx_0 = mg$$

or

$$x_0 = \frac{g}{k}m \tag{2}$$

which is a linear relation in x_0 versus m. Today's experiment will attempt to verify Hookes' law in the form of Eq. (2) by measuring the variation of static displacement x_0 with respect to suspended mass m for a spring-mass system. From these results, the spring constant k will also be calculated.

If the mass is pulled a small distance x' below its equilibrium position (Fig. 1) and released, the mass will execute simple harmonic motion (SHM) about the equilibrium position. The distance x', which is the maximum displacement of the mass on either side of the equilibrium position, is called the *amplitude* of the SHM. One complete oscillation is called a *cycle* and the time required for 1 cycle is called the *period T*, and is measured in seconds. The period is the time taken by the spring to go from its maximum extension x', through its maximum compression $-x'$, and back again to its maximum extension. The number of cycles completed per second is called the *frequency f*, and is measured in cycles per second (sec^{-1}). (The frequency is the reciprocal of the period: $f = 1/T$). For the SHM which results from a mass m attached to a spring of spring constant k, the period is given by

$$T = 2\pi\sqrt{\frac{m}{k}} \tag{3}$$

It should be noted that for small amplitudes of vibration (be careful never to stretch the spring beyond its elastic limit!), the period does not depend upon the amplitude x': The same period results no matter how much or how little the mass is initially displaced within its elastic limit.

Squaring Eq. (3) gives

$$T^2 = 4\pi^2 \frac{m}{k}$$

which, with Eq. (2), can be written as

$$T^2 = \frac{4\pi^2}{g} x_0 \tag{4}$$

which is a linear relation in T^2 versus x_0 with a theoretical slope of $4\pi^2/g$. This relation will be tested experimentally by determining the variation in the period T with respect to static displacement x_0 (i.e., for various suspended masses m). Moreover, the calculated period using Eq. (3) will be compared with the observed period.

Procedure and Analysis of Data

1. Hang the spiral spring from the support. With the meter stick vertically alongside the spring, record the position of the lower end of the spring.
2. Suspend a *total* mass (including weight hanger) of 500 g from the spring and again record the position of its lower end, from which the static displacement x_0 may be obtained.
3. Pull the mass a small distance (less than $x_0/2$) below its equilibrium position and release. Measure the time taken for the mass to complete 50 oscillations or cycles. Reduce the number of cycles if the system is rapidly "damped" (decrease in amplitude as energy of system is lost due to friction). Record the total time and number of cycles. Calculate the average period T by dividing the total time by the number of cycles.
4. Repeat steps 2 and 3 with total suspended masses of 750, 1000, 1250 and 1500 g.
5. Make a plot of static displacement x_0 versus total suspended mass m. Calculate the value of the slope from your graph and equate to g/k from Eq. (2) to determine the spring constant k of your spring.
6. Make a plot of the period squared T^2 versus static displacement x_0. Calculate the value of the slope from your graph and compute the percent error between it and the theoretical value $4\pi^2/g$ from Eq. (4).
7. Using your experimentally determined value of k and Eq. (3), calculate the value of the period T for the 1500-g mass. Compare this calculated value of the period with the corresponding observed value by computing the percent difference.
8. To show that the period is not dependent on amplitude, remount the 1000-g mass. Determine the periods obtained when three different displacements are used (say 2, 4, and 6 cm). The results should be identical (to within the measurement error).

EXPERIMENT 5

SIMPLE HARMONIC MOTION
REPORT SHEET

Name _____   Section _____

Total Suspended Mass m, g	Lower End of Spring Scale Reading, cm	Static Displacement x_0, cm	Total Time, sec	No. of Cycles	Period T, s	Period Squared T^2, s^2
0		XXX	XXX	XXX	XXX	XXX
500						
750						
1,000						
1,250						
1,500						

For part 8:

Displacement (cm)	Period determined (T, s)
(1) _____	(1) _____
(2) _____	(2) _____
(3) _____	(3) _____

1. Have you verified Hooke's law for your spiral spring? Explain.

2. (a) Slope of plot of x_0 versus m: _____   (b) Spring constant $k = $ _____

3. Have you verified the period relationship for SHM, Eq. (3), for your spring-mass system? Explain.

4. (a) Slope of plot of T^2 versus x_0:   (b) Percent error between slope and $4\pi^2/g$:

   _____   _____

5. (a) Calculated value of period for $m = 1500$ g: $T = $ _____

   (b) Percent difference with observed value: _____

6. What is the frequency of the SHM when $m = 500$ g? $f =$ _____ [Use Eq. (3) and your experimentally determined value of k.]

7. Compare the results found for step 8 of the procedure. Are the periods essentially identical (to within the error found in 5 above)? Is the period independent of the amplitude? Why?

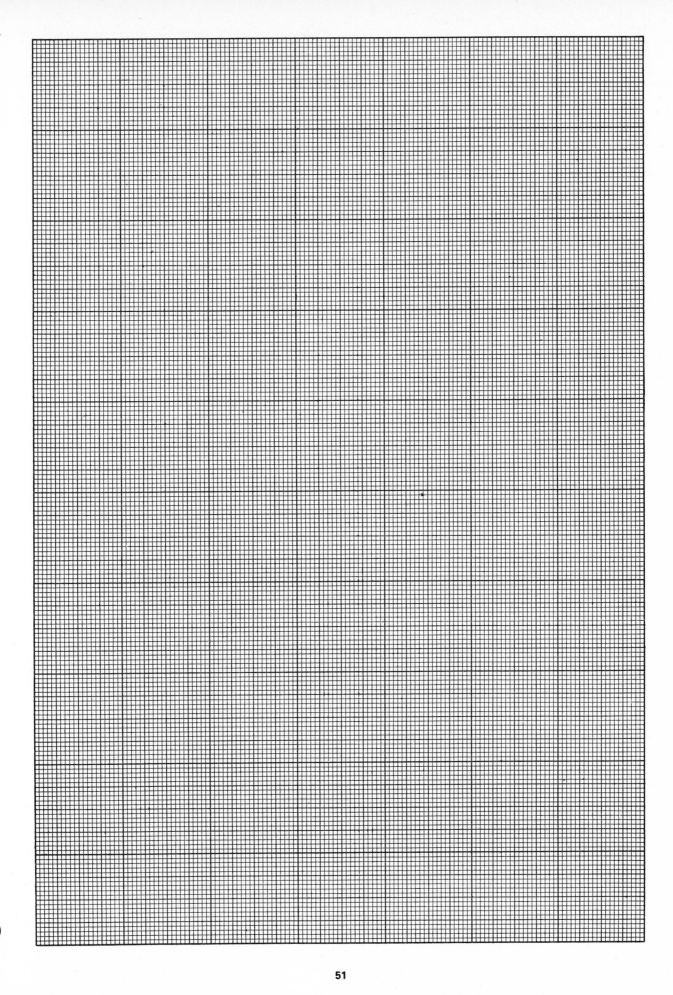

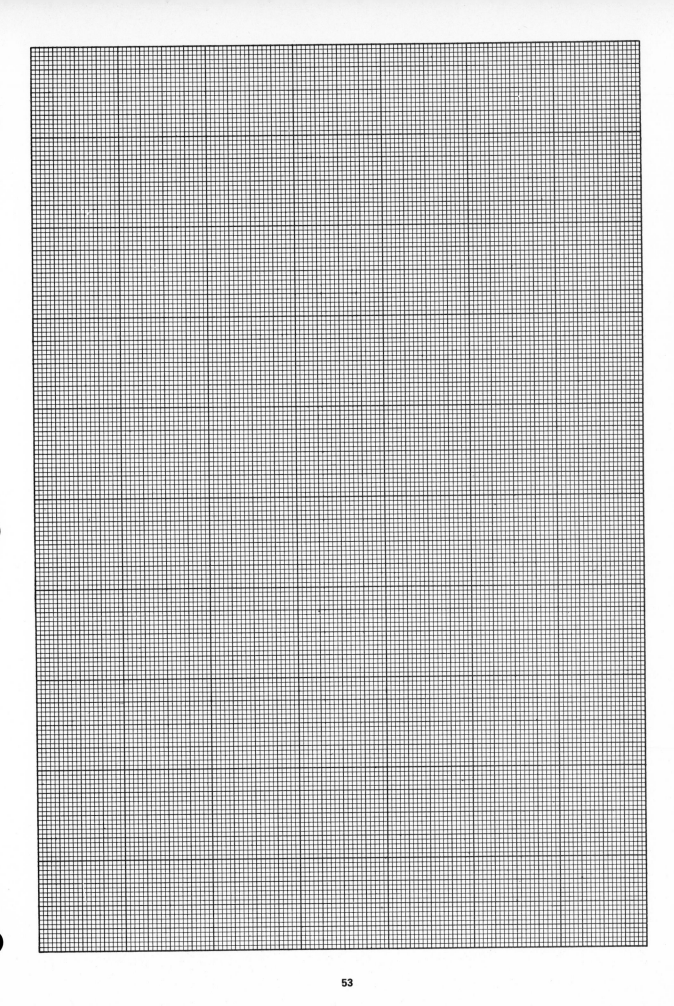

EXPERIMENT 6

THE SIMPLE PENDULUM

Materials Needed

String
Three pendulum bobs of different masses
Rigid support
Meter stick
Protractor (if available)
Laboratory timer or stopwatch

Background

The scientific method states that no scientific theory or model of nature is acceptable unless the results it predicts agree with experiment. In essence, scientists try to theoretically *predict* physical phenomena and the behavior of physical systems, then *test* their theories against experiments in the laboratory. The theory is considered to be a valid and accurate description of the physical phenomena only if there is strong agreement between the experimental results and the theoretical predictions.

The present experiment will illustrate the use of the scientific method by investigating the motion of a simple pendulum. You will be given a theoretical expression or equation that describes the behavior of a simple pendulum, and you will test the validity of this theoretical relationship experimentally. In the process, you will study the factors which determine the period of a simple pendulum. Moreover, you will use your results to make an experimental determination of the acceleration due to gravity and to study the behavior of other pendulums of different properties.

The simple pendulum consists of a small concentrated mass, or "bob," suspended by a cord of negligible mass.

The physical parameters* involved in the motion of a simple pendulum are:

1. The *length L,* measured from the point of suspension to the center of the pendulum bob.
2. The *mass m* of the pendulum bob.

*Parameter: one of a set of physical quantities or properties whose values determine the characteristics or behavior of a system.

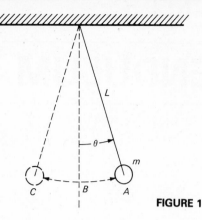

FIGURE 1

3. The *angular displacement* θ through which the pendulum swings (also called the *amplitude*), which is the maximum displacement (*A* or *C* in Fig. 1) from the equilibrium position (*B* in Fig. 1).

A phenomenon which repeats itself in a regular fashion is said to be *periodic*. If the pendulum bob is displaced to one side and released, periodic motion results. The period *T* of the pendulum is the time it takes for the pendulum to swing through one complete oscillation or cycle of its motion (e.g., from *A*—the point of release—to *C* and back to *A* in Fig. 1).

If the pendulum bob is allowed to swing through a *small* arc or angular displacement, θ (less than 10°), then the theoretical expression for the period *T* of the pendulum is

$$T = 2\pi \sqrt{\frac{L}{g}} \tag{1}$$

where *g* is the acceleration due to gravity and has an accepted value of $g = 980$ cm/s^2. This equation is derived from considering the fact that for small angular displacements the restoring force on the pendulum bob is proportional to the negative displacement. Since this is the condition for simple harmonic motion, the pendulum will oscillate accordingly. For larger displacements, the equation which describes the motion is much more complicated and no longer has such a simple result. The motion, then, is no longer harmonic. Be very careful to use ony small angular displacements (similar to those suggested in this lab). It must be noted that in this small angle expression for period neither the angular displacement θ nor the mass *m* of the bob appears. From this equation, then, what factors or parameters determine the period of the pendulum? Is the period theoretically independent of any of the pendulum parameters?

By making measurements of *L* and *T*, the relationship expressed by Eq. (1) will both be tested experimentally and used to determine the acceleration *g* due to gravity.

General Procedure. The procedure used by the scientist in the laboratory to study the behavior of a system is to choose some property of the system and perform experiments to determine how this property depends on the other physical properties of the system. The property of interest here is the period *T* of a simple pendulum. Experiments must be performed, then, to determine how the period *T* depends upon the other properties of a simple pendulum such as its length *L* and its mass *m*. This is accomplished by varying one of these properties, length, for example, while keeping all the other parameters or properties the same and measuring the corresponding period each time. Next, a different property is varied, the mass of the bob, again keeping all other parameters constant and determining how this affects the period. By this method, the behavior of the system is revealed and the theoretical relationship for the period *T* is tested.

Procedure and Analysis of Data

1. Make a simple pendulum arrangement using a length L of 150 cm of string and select one of the bobs or masses m. Make sure that the string is secure and does not slip on the support.

2. Displace the pendulum bob to one side through an angle of not more than 10° and let the pendulum oscillate. Measure the amount of time taken for 20 complete swings or cycles, t_{20}. In counting cycles, be sure to start the stopwatch on the count of zero. (To minimize timing error, it is also best to start or stop the clock as the bob passes through the middle position.) Calculate the corresponding experimental value of the period T, the time for 1 cycle, by dividing the total time for 20 cycles, t_{20}, by 20. Record L, t_{20}, and T in Table 1 of your Report Sheet.

3. Obtain the square of the period, T^2, and also, using Eq. (1), calculate the theoretical period T' for this pendulum length. Record these values in Table 1.

4. To experimentally investigate the relationship between the period T and length L, repeat parts 1, 2, and 3 making the length of pendulum successively 120, 90, 60, and 30 cm while keeping the mass constant. (**Caution**: As the string is shortened, be careful to keep the angular displacement within the 10° limit specified.) Record all data in Table 1.

5. Make a plot of period squared (T^2) versus length L (use the first graph at the end of this experiment). (Read the instructions on plotting graphs in the Introduction.)

6. To calculate g, both sides of Eq. (1) are squared, giving:

$$T^2 = \left(\frac{4\pi^2}{g}\right) L \qquad (2)$$

This equation has the form $y^2 = bx$, that of a parabola. But as was discussed in the last section of the Introduction, this is made to be a linear relation with a change of the dependent variable from T to T^2. Calculate the value of the slope from your graph and record on your Report Sheet. An expression for the slope, m, of this straight line is also obtainable from Eq. (2) above as:

$$m = \frac{4\pi^2}{g}$$

Equating the value of the slope from your graph with the above expression for the slope, Eq. (3), determine an experimental value for g and record on the Report Sheet. Compute the percent error between the experimental and accepted values of g.

7. To experimentally investigate the relationship between the period T and mass m of the pendulum bob, repeat steps 1 and 2, this time varying the mass three times but keeping the length constant at some convenient length. Measure the mass of each bob, with the pan balance. Record m, t_{20}, and T in Table 2 of your Report Sheet.

EXPERIMENT 6

THE SIMPLE PENDULUM
REPORT SHEET

Name _____   Section _____

Table 1   period T versus length L; pendulum mass m = _____

Trial	1 Length L, cm	2 Time of 20 cycles, t_{20}, s	3 Experimental Period T, s	4 Square of Period T^2, s^2	5 Theroretical Period, T', s
1.					
2.					
3.					
4.					
5.					

Table 2   Period T versus mass m; pendulum length L = _____

Trial	1 Mass m, g	2 Time of 20 cycles, t_{20}, s	3 Experimental Period T, s
1.			
2.			
3.			

(For the following questions show all necessary calculations on additional pages.)

1. What parameters does the period of a simple pendulum depend upon and what parameters is it independent of?

2. One of the objects of the preceding experimental procedures was to determine the validity of Eq. (1), that is, whether the experimental results agreed with the theoretical predictions as required by the scientific method. Have we shown the validity of Eq. (1) and how? *Hint*: Compare columns 3 and 5 of Table 1 in the Report Sheet and how well your points and the best fit line correspond on your graph.)

3. Calculated slope of the straight line from the graph: _____

4. (Experimental) equation of the line: _____

5. Experimental value of g, g_{exp} = _____

6. Percent error $|(g - g_{exp})/g| \times 100$ = _____

7. Using your graph and/or your equation obtained in question 4, determine the length of a Foucault pendulum which has a period T of 7.66 s. L = _____

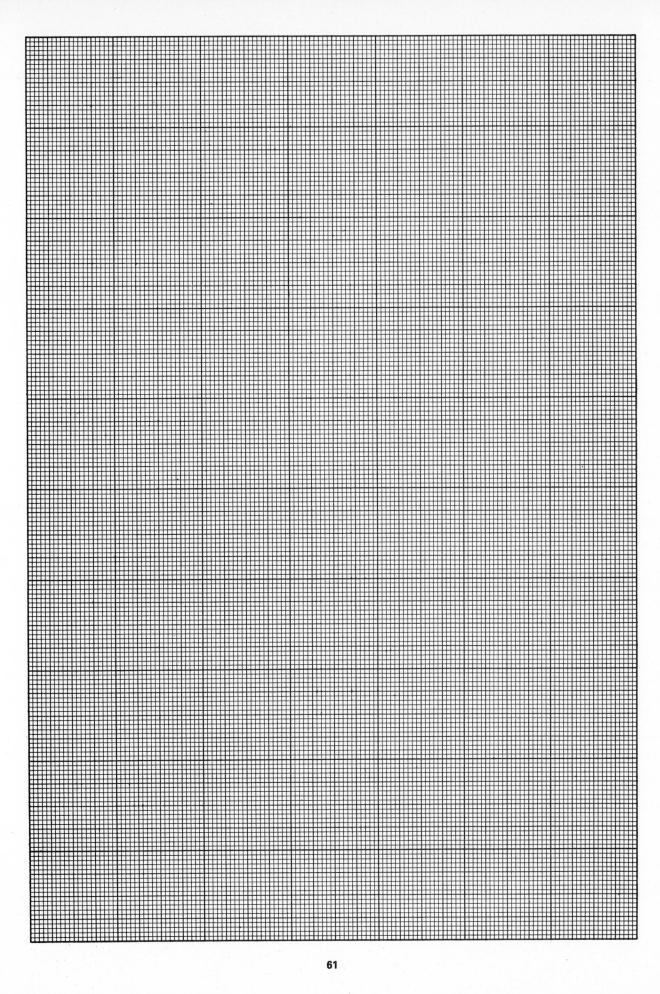

EXPERIMENT 7

THE BALLISTIC PENDULUM

Materials Needed

Ballistic pendulum apparatus
Meter stick
Laboratory Balance

Background

The ballistic pendulum is a device commonly used to measure the velocity of a projectile. It is interesting and illustrative as a laboratory activity because it shows two conservation principles at work. These principles are conservation of *momentum* and conservation of *energy*. The principle of conservation of energy may be stated: Energy is neither created nor destroyed, but may be converted from one form to another. Of course when adding up the total energy of some system, we must be sure that we count *all* forms of energy.

The two forms of energy important when studying a pendulum are *kinetic* and *potential* energy. Kinetic energy (KE) is energy possessed because of motion and is given by KE = $\frac{1}{2}$ mv^2. Potential energy (PE) is energy possessed because of position (in the earth's gravitational field) and is given by PE = mgh, where h is the height above the level where the PE is zero. The choice of the level zero PE is *arbitrary* and may be chosen for convenience in working a problem.

The ballistic pendulum is a large mass (called the pendulum bob) supported by string or very light rod. See Fig. 1a. The projectile is allowed to imbed itself in the bob, causing the bob to have kinetic energy. See Fig. 1b. The bob then swings upward, converting the kinetic energy to potential energy. At the highest point of its swing, all of the kinetic energy has been converted to potential energy. See Fig. 1c. A ratchet is usually employed to stop the pendulum at the highest point so that h can be measured.

Now conservation of energy requires that the kinetic energy at the bottom of the pendulum's swing must equal the potential energy at the top of the swing, that is

KE (bottom) = PE (top)

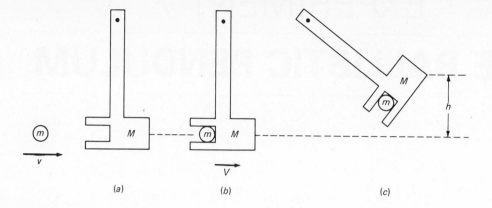

(a)   (b)   (c)

or

$$\tfrac{1}{2} (M + m) V^2 = (m + M) gh$$

We use $m + M$ because it is the total mass of the system. If h is measured, then this last equation may be used to determine V. Note that the mass $m + M$ cancels out! The expression for V is

$$V = \sqrt{2gh}$$

Now that we can obtain V from a measurement of h, we turn to conservation of momentum in the collision (inelastic) between the projectile and the bob. Initially, *only* the projectile is moving and hence possesses momentum. After collision the projectile and bob stick together (energy not conserved) and move with a common velocity V. (Remember momentum is the product mass × velocity.) Conservation of momentum requires

Momentum before collision = momentum after collision

or

$$mv + M(0) = (m + M)V$$
$$mv = (m + M)V$$

Therefore, if the masses m and M are known and V has been calculated, we may obtain the velocity of the projectile v from the last equation above:

$$v = \frac{(m + M)V}{m}$$

A measurement of the height to which the pendulum swings, coupled with the principles of conservation of energy and momentum, allows the determination of the velocity of the projectile. Applications of this range from determining the muzzle velocity of a gun to measuring the speed that a person can throw a baseball.

If the projectile had missed the bob, it would have described a curved path to the floor. By measuring the height from which it fell and the distance it traveled in the air, its initial velocity could have been determined.

Procedure

1. Get the gun ready for firing. Release the pendulum from the rack and allow it to hang freely. When the pendulum is at rest, pull the trigger, thereby firing the ball into the pendulum bob. This will cause the pendulum with the ball inside it to swing up along the rack where it will be caught at its highest point. Record the notch on the curved scale reached by the pawl when it catches the pendulum. To remove the ball from the pendulum, push it out with the finger or a rubber-tipped pencil, meanwhile holding up the spring catch.

2. Repeat procedure 1 four more times, recording the position of the pendulum on the rack each time on the Report Sheet.

3. From the data of procedures 1 and 2, compute the average value of the position of the pendulum on the rack. Set the pendulum with the pawl engaged in the notch which corresponds most closely to the average reading. Measure the vertical distance h_1 from the base of the apparatus to the index point attached to the pendulum. This index point indicates the height of the center of gravity of the pendulum and ball. Use the metric steel scale and read the measurement of 0.1 mm. Record distance h_1 on the Report Sheet.

4. With the pendulum hanging in its lowest position, measure the vertical distance h_2 from the base of the apparatus to the index point. Record this value on the Report Sheet. Record the vertical distance h through which the center of gravity of the pendulum was raised, $h_1 - h_2$ on the Report Sheet.

5. Compute the value of V, the common velocity of the pendulum bob and ball just after the collision, by using the relation

$$V = \sqrt{2gh}$$

Record this value on the Report Sheet.

6. Loosen the thumbscrew holding the axis of rotation of the pendulum and carefully remove the pendulum from its support. Weigh the pendulum and the ball separately and record the values obtained on the Report Sheet.

7. Calculate the velocity of the ball before the collision by substituting the value of V found in calculation 5 and the measured values of the masses of the pendulum bob and of the ball in the momentum equation. Record this value on the Report Sheet.

8. In the second part of the experiment, the initial velocity of the projectile is obtained from measurements of the range and fall. The apparatus should be set near one edge of a level table. In this part the pendulum is not used and should be swung up onto the rack so that it will not interfere with the free flight of the ball.

9. Get the gun ready for firing by placing the ball on the end of the firing rod and pushing it back, compressing the spring until the trigger is engaged. The ball is fired horizontally so that it strikes a target placed on the floor. Fire the ball and determine approximately where it strikes the floor. Place a sheet of white paper on the floor so that the ball will hit it near its center and cover it with carbon paper; place some weights at the corners of the paper to keep it from moving around. In this way a record can be obtained of the exact spot where the ball strikes the floor. Fire the ball five more times.

10. Measure the range of each shot; this is the horizontal distance from the point of projection to the point of contact with the floor. Use a plumb bob to locate the point on the floor directly below the ball as it leaves the gun. Record the values of the five measurements on the Report Sheet. Calculate the average and record it on the Report Sheet.

11. Measure the vertical fall of the ball, that is, the vertical distance of the point of projection above the floor. Record this value on the Report Sheet.

12. From the measured vertical distance that the ball falls and the known value of g, calculate the time of flight of the ball, by using the equation

$$t = \sqrt{\frac{2y}{g}}$$

Record this result on the Report Sheet.

13. Compute the velocity of projection from the time of flight obtained in calculation 12 and the range obtained in calculation 10, using $v = x/t$. Record this on the Report Sheet.

14. Find the percent difference between the values of the velocity of the ball in steps 7 and 13.

EXPERIMENT 7

THE BALLISTIC PENDULUM
REPORT SHEET

Name _____   Section _____

3. Rack reading

Trial	1	2	3	4	5	Average

4. $h_1 = $ _____ cm

   $h_2 = $ _____ cm

   $h = h_1 - h_2 = $ _____ cm

5. $V = \sqrt{2gh} = $ _____ cm/s velocity of bob

6. $m_{bob} = $ _____ g

   $m_{ball} = $ _____ g

7. $V = $ _____ cm/s velocity of ball (muzzle velocity)

$$v = \frac{(m + M)}{m} V$$

10. Range of projectile (cm) ×

Trial	1	2	3	4	5	Average

11. Vertical distance $y = $ _____ cm

12. Time of fall $t = \sqrt{2y/g} = $ _____ s

13. Velocity of projectile $v = x/t = $ _____ cm/s

14. % difference $= (v - v')/$average $\times\ 100 = $ _____

15. Which v do you consider more accurate? _____

    Why? _____

    _____

16. What two conservation principles were involved in this experiment?

17. (Optional.) Calculate the fractional loss of kinetic energy:

$$\frac{KE_{(initial)} - KE_{(final)}}{KE_{(initial)}}$$

where

$KE_{(initial)} = \frac{1}{2} mv^2$

$KE_{(final)} = \frac{1}{2} (m + M) V^2$

Actual KE loss = _____ (use the v you consider to be most accurate)

Theoretical KE loss = $\dfrac{M}{m + M}$ = _____

% error = _____ (between actual KE loss and theoretical KE loss)

What happened to the missing energy? _____

EXPERIMENT 8

ARCHIMEDES' PRINCIPLE

Materials Needed

Laboratory balance
Irregular- and regular-shaped masses (a solid aluminum cylinder is referred to in this
   lab, but other regular shapes may of course be used)
String and paper clips
Large cup with draining tube
Small cup to catch displaced water
Paper towels

Background

Anyone swimming or boating has directly experienced the effects of buoyancy. Perhaps while swimming you have attempted to lift a heavy object out of the water and experienced the sudden increase in (apparent) weight of the object just as it starts to come out of the water. Or in boating, have you noted the boat getting lower and lower in the water as its load is increased?

Archimedes first put the ideas of buoyancy on a quantitative basis. Simply stated his principle is: An object immersed in a liquid is *buoyed* up by a force equal to the *weight* of the liquid displaced. Mathematically, Archimedes' principle may be stated as $F_B = W_L$, where F_B is the buoyant force and W_L is the weight of the liquid displaced. Now the weight of a given volume of liquid is $W_L = \Delta mg$, where Δm is the weight of the liquid displaced.

The volume of liquid displaced will be equal to the volume of the immersed object, provided the object does not collapse. The principle of buoyancy can be used to do many things. It can be used to determine the density of a liquid, the density of a solid, or the volume of an irregular object. All of these will be done in this laboratory. *Mass* and *weight* will again be important, so the student should review mass and weight and how to convert from one to the other.

Archimedes' principle can be used to determine the mass density of liquids and solids and to determine the volume of an irregular object. All of these depend on measuring the weight of an object in air, and then again with the object immersed in the liquid.

If a balance is used that determines mass, we will often talk about "apparent mass."

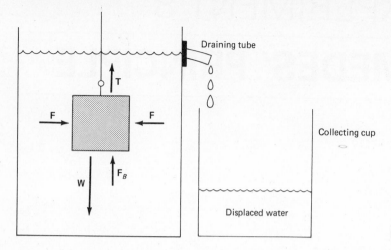

Draining tube

Collecting cup

Displaced water

FIGURE 1

Figure 1 shows the forces acting on an object immersed in a liquid. The forces on the sides are equal and opposite, so they cancel out. The tension T in the supporting string is what the balance will measure as weight or apparent mass. The weight W of the object acts down while the buoyant force F_B and the tension T act upward. If the object sinks (F_B less than W), we will record T as the apparent weight of the object, whereas if the object floats (F_B greater than W), we will record zero for T. In this case, however, we simply attach a weight to the object in order to sink it.

When the object hangs at rest in the water, we know that the sum of the upward forces must equal the sum of the downward forces and we may write that $T + F_B = W$. T will measured on the balance to which the string is attached. The balance (Figure 2) can be used to measure W also—the object is simply raised out of the liquid. The buoyant force will be simply the difference between the weight of the liquid and the weight immersed in the liquid:

$$F_B = W - T$$

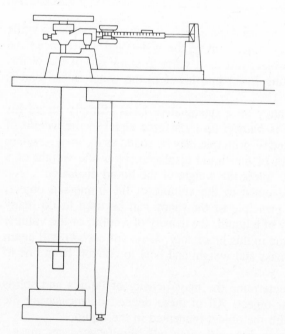

ONE SETUP FOR ARCHIMEDES' PRINCIPLE   **FIGURE 2**

Remember that the volume of the immersed object is equal to the volume of the displaced liquid (water). This is found from the liquid's density, ρ_w, and the mass displaced:

$$V = \frac{\Delta m}{\rho_w}$$

Procedure and Analysis of Data

1. Determine the mass of the irregular object by suspending the mass from the pan balance in the air (m_a).
2. Fill the large cup with water up to the bottom of the draining tube. "Wiggle" the tube with the cup resting on a flat surface so that all the excess water is removed. Weigh the small cup (m_c). Now place the small cup in front of the spout (draining tube) and lower the suspended object into the water until it is completely submerged. Determine the mass of the object in water (m_w). Also determine the mass of the small cup with water (m_{cw}).
3. From this information determine
   (a) The mass of the displaced water $= m_{cw} - m_c = \Delta m$.
   (b) The volume of the displaced water $= V = \Delta m/\rho_w$ where ρ_w is the density of the water (1 g/cm^3). This is equal to the volume of the irregular object.
4. Determine the density (ρ_0) of the irregular object using the mass and volume previously determined.
5. Now compute the buoyant force: $\mathbf{F}_B$ = weight of object in air − weight of object in water or, $\mathbf{F}_B = m_a g - m_w g$, where $g = 980$ cm/s^2.
6. Compare this with the weight of the displaced fluid (Δmg) by computing the percentage difference:

$$\% \text{ difference} = \frac{|\mathbf{F}_B - \Delta mg|}{(\mathbf{F}_B + \Delta mg)/2} \times 100$$

7. Repeat the same procedure with the aluminum cylinder but also determine the volume of the cylinder analytically (by measuring with a vernier caliper). Also compute a percentage difference between the "analytical volume" and the volume determined by displacement of the water (volume $= \pi r^2 h$, where r is the radius of the cylinder and h is the height of the cylinder). Enter this information in Table 2 of your Report Sheet. Answer the questions.

EXPERIMENT 8

ARCHIMEDES' PRINCIPLE
REPORT SHEET

Name _____   Section _____

Table 1   Irregular object

Mass of object (air) m_a =	
Mass of object (water) m_w =	
Mass of empty cup m_c =	
Mass of filled cup m_{cw} =	
Mass of displaced water =	
Volume of object v =	
Density of object ρ_0 =	
Buoyant force $\mathbf{F}_B$ =	
Weight of displaced water =	
Percentage difference =	

Table 2   Cylinder

Mass of the cylinder in air =	
Mass of the cylinder in water =	
Mass of the displaced water =	
Analytic volume =	
Displaced volume =	
Percentage difference =	
Density of cylinder (use average volume) =	
Bouyant force =	
Weight of displaced water =	
Percentage difference =	

1. In Table 2, why should the analytic volume and the displaced volume of the cylinder be equal?

2. Have you confirmed Archimedes' principle? Why?

3. (Optional.) How could you determine the density of an object simply by weighing it in water and in the air? This density determination is called *specific gravity*. Determine the density of the irregular object by this means. How does this value of density compare with that determined in Table 1? Find the percentage difference.

Explanation: _____

_____

Specific gravity of object _____

Percentage difference _____

EXPERIMENT 9

BOYLE'S LAW

Materials Needed

Boyle's law apparatus

Background

It is common knowledge that when a fixed amount of gas is subjected to pressure, its volume decreases. Exact calculations of the volume-pressure relationship were performed by Robert Boyle, among others, who in 1662 concluded his studies by saying that at a fixed temperature and with a fixed amount of gas, the volume is inversely proportional to the pressure, $V \propto 1/P$. This observation may be restated as $V_1/V_2 = P_2/P_1$. As the volume of gas increases, its pressure decreases. The proportion $V \propto 1/P$ may be made an equation by introducing a constant k:

$$V = \frac{k}{P}$$

It is important to note that the product of volume $\times$ pressure is always a constant: $V \times P = k$.

As an analogy, consider the following proportions:

$$2 \propto \tfrac{1}{4} \quad 4 \propto \tfrac{1}{2} \quad 8 \propto \tfrac{1}{1}$$

These proportions all have the same constant, $k = 8$:

$$2 = \frac{k}{4} \quad 4 = \frac{k}{2} \quad 8 = \frac{k}{1} \quad \text{where } k = 8$$

The products

$$2 \times 4 = k \quad 4 \times 2 = k \quad 8 \times 1 = k \quad \text{all give 8}$$

In this experiment, you will repeat the experiment that led Boyle to his conclusion.

Procedure

(Caution: Mercury and mercury vapors can be harmful.) Your instructor will demonstrate the use of the instrument. Work in groups of two to four students per apparatus, taking a total of eight readings.

The following instructions apply to the instrument shown in Fig. 1. Other instruments may require slightly different instructions. *The apparatus is fragile and contains mercury and must therefore be handled with extreme care.* Be sure you follow instructions explicitly. The apparatus shown in Fig. 1 has two columns. The column with the stopcock will be referred to as A and the other as B.

Move column A toward the top of the scale and fasten tightly. Record the position of the juncture between the stopcock and the tube as "Stopcock position" on the Report Sheet. This reading should be made in millimeters. Leave column A fixed in this position for the remainder of the experiment.

Now adjust the height of column B such that the mercury level is about the same in columns A and B and open the stopcock in A. With the stopcock open, read and record the exact position of the levels in A and B.

Close the stopcock, raise column B somewhat, and again record the precise levels of A and B. Raise the levels two or three more times, each time recording both levels and always keeping the stopcock closed. Then lower column B a few times below the initial starting line

FIGURE 1   An apparatus used to observe Boyle's law. *(Courtesy of W. K. Fife.)*

again recording levels A and B each time. Be careful that the tubing containing the mercury is allowed to hang freely at all times without touching the floor.

Record the atmospheric pressure of the day. Recall that 760 mmHg = 760 torr = 1 atm. Now proceed to calculate $B - A$, the pressure differential experienced by the air in column A; atmospheric pressure + $(B - A)$, which is the total pressure experienced by the air in column A; stopcock position $- A$, which gives the volume changes; and finally obtain the product of pressure × volume. This last column should give values that are fairly constant if Boyle's law was indeed observed. Carry out these calculations *before* leaving the laboratory.

EXPERIMENT 9

BOYLE'S LAW
PRE-LABORATORY QUESTIONS

1. State the relationship between pressure and volume as stated by Robert Boyle. Remember to state the condition of temperature.

2. A sample of gas has a pressure of 155 torr while occupying a 100-mL flask at 34°C. If the sample is transferred into a 500-mL flask at 34°C, what is the pressure due to the sample of gas?

EXPERIMENT 9

BOYLE'S LAW
REPORT SHEET

Name _____   Section _____

I. DATA

Atmospheric pressure _____ mmHg   Stopcock position _____ mm

	A	**B**	**Difference*** **B–A**	**Pressure** **P**	**Volume†** **V**	**Product‡** **PV**
With Stopcock Open						$\times\ 10^5$
						$\times\ 10^5$
						$\times\ 10^5$
						$\times\ 10^5$
With Stopcock Closed						$\times\ 10^5$
						$\times\ 10^5$
						$\times\ 10^5$
						$\times\ 10^5$
						$\times\ 10^5$

A = Hg level in column A; B = Hg level in column B; P = atmospheric pressure + $(B - A)$; V = stopcock position − A.
*Some readings $(B - A)$ will be negative.
†The length of the tube is proportional to its volume. The length reading is taken simply because it is more accurate with this instrument.
‡Be careful with significant figures!

II. CALCULATIONS

1. Plot the pressure versus volume onto the graph paper on the next page. Label the *left-hand* vertical axis "Pressure" and the bottom horizontal axis "Volume." Be sure to *expand* the

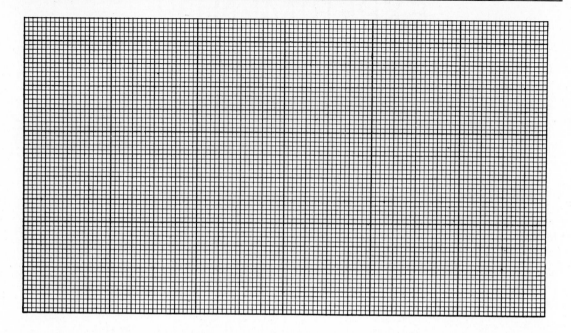

values as much as possible, using the entire graph grid for your plot (do not crowd all your data into one corner of the graph).

2. Onto the graph provided, but using a different color pen, plot the pressure × volume product versus the volume. To do so, calibrate the *right-hand* side vertical axis from zero at the bottom to 2.0×10^5 at the top. Leave the horizontal axis as in 1.

3. Is Boyle's law confirmed by your graph? Explain.

Problems

1. A 300-mL cylinder contains a gas at 780 mmHg. What is the pressure when the cylinder is compressed down to 40 mL?

_____ mmHg

2. As a balloon rises into the atmosphere, the pressure around the balloon decreases, and as a result the balloon expands. Assume a perfectly expansible balloon inflated at sea level to 100 liters. The maximum volume it can withstand before bursting is 350 liters. At what height will the balloon burst? (See Table 1).

_____ km

Table 1   Atmospheric pressure versus altitude (approximate)

Altitude above Sea Level, km	Pressure, mmHg
0	760
2	600
5	400
10	210
13	150

EXPERIMENT 10

CALORIMETRY: SPECIFIC HEAT AND LATENT HEAT OF FUSION

Materials Needed

Calorimeter
Stove and bucket
Metal samples with attached string
Two thermometers
Laboratory balance
Alcohol or Bunsen burner

Background

Heat is that form of energy which is transferred from one body to another by virtue of a temperature difference between them. Heat spontaneously flows from the hotter to the colder body. Provided no work is done by or on a body, when the body gains or loses heat energy it will either change its temperature or its phase (solid, liquid, or gas). The purpose of this experiment will be to study these two processes using the experimental technique of calorimetry, which enables the quantitative measurement of heat exchange. The specific heat of a metal and the latent heat of fusion of water will be obtained.

I. DETERMINATION OF SPECIFIC HEAT OF A METAL

When heat is absorbed, the temperature of a body usually increases. Different substances, however, require different quantities of heat, Q, to produce a given temperature change ΔT. The response of a particular material to heat is expressed quantitatively in terms of the *specific heat c* of the material. Specific heat is defined as the heat energy absorbed per unit mass m, per degree rise in temperature, i.e.,

$$c = \frac{Q}{m\Delta T} \tag{1}$$

The specific heat is the amount of heat (in calories) required to change the temperature of 1 g of a substance 1°C. The specific heat of water, for example, is 1.00 cal/g·°C. Equation (1) implies that if the temperature of a body changes by ΔT, the heat Q that flows into or out of it is given by

$$Q = mc\Delta T \tag{2}$$

From the definition, the specific heat of a material can be determined experimentally by measuring the temperature change of a known mass of a material produced by adding a known quantity of heat. This determination is easily accomplished indirectly by a calorimetry method known as the *method of mixtures*. It is based on the principle of conservation of (heat) energy. When two or more substances at different temperatures are placed together in an isolated system (no work is done on or by the system and no heat is lost to or gained from the surroundings), the warmer substances will lose heat and the colder substances will gain heat until everything reaches a common equilibrium temperature. Within the system, the total amount of heat lost by the warmer substances must equal the total amount of heat gained by the colder substances, or

$$\text{Heat lost } = \text{ heat gained} \tag{3}$$

In part 1 of this lab, a hot metal sample is put into cool water in a calorimeter cup. The calorimeter insulates the system from heat losses. After stirring, the water-metal-calorimeter combination reaches an intermediate equilibrium temperature. Using Eqs. (2) and (3):

Heat lost by sample = heat gained by water × heat gained by calorimeter

$$m_s c_s \Delta T_s = m_w c_w \Delta T_w + m_{cal} c_{cal} \Delta T_{cal}$$

$$m_s c_s (T_h - T_e) = m_w c_w (T_e - T_c) + m_{cal} c_{cal} (T_e - T_c) \tag{4}$$

where T_h is the initial temperature of the hot metal sample, T_c the initial temperature of the colder water and calorimeter cup, and T_e the final intermediate equilibrium temperature of the system. Equation (4) may be used to determine the specific heat of the metal sample, c_s, as all other quantities are either known or measurable.

II. LATENT HEAT OF FUSION OF WATER

The addition or removal of heat from a substance does not always lead to a change in its temperature. Instead the substance may change its phase at a constant temperate. The change of phase may be from solid to liquid (fusion) or from liquid to gas (vaporization) when heat is added, or it may change from gas to liquid (condensation) or from liquid to solid (solidification) when heat is removed. The heat that is absorbed or released during this change of phase, at constant temperature, is known as *latent heat l*. It is defined as the amount of heat required to change the phase of a unit mass of a substance, at a constant temperature, and has the units of cal

g. The *latent heat of fusion l_F* of a substance, for example, is the heat per unit mass required to change the substance from the solid to the liquid phase at its melting temperature. It also refers to the heat per unit mass released when the substance changes from a liquid to a solid. Mathematically,

$$l_F = \frac{Q}{m} \tag{5}$$

The latent heat of fusion of water, for example, is 80 cal/g: 80 cal of heat are needed to convert 1 g of ice into 1 g of water at its melting point of 0°C.

In part 2 of today's lab, the latent heat of fusion of water will be found by the method of mixtures. A quantity of ice of mass, m_i, at 0°C is added to warm water in a calorimeter cup at temperature T_h. After stirring, the ice melts and the water-calorimeter combination reaches an intermediate equilibrium temperature T_e. Then by conservation of energy for the mixture, Eq. (3), which holds even if a change of phase occurs:

Heat lost by warm water + heat lost by calorimeter = heat of fusion to melt ice
+ heat gained by ice water

or

$$m_w c_w \Delta T_w + m_{cal} c_{cal} \Delta T_{cal} = m_i l_F + m_i c_w \Delta T_{iw}$$

$$m_w c_w (T_h - T_e) + m_{cal} c_{cal} (T_h - T_e) = m_i l_F + m_i c_w (T_e - 0) \tag{6}$$

Equation (6) may be used to determine the latent heat of fusion of water, l_F, as all other quantities are either known or measurable.

Procedure and Analysis of Data

I. DETERMINATION OF THE SPECIFIC HEAT OF A METAL

1. Fill the bucket half full of water and drop in the metal sample(s). Put a thermometer in the water and start heating the water on the stove. (Keep the thermometer bulb off the bottom of the bucket.)
2. Determine the mass of the inner calorimeter cup, m_{cal}. Also note the specific heat of the cup, c_{cal}, which is usually stamped on the cup.
3. Fill the calorimeter cup two-thirds full with cold tap water and determine the mass of the cup and water from which the mass of water, m_w, can be obtained. Place the calorimeter cup with the water in the calorimeter jacket and put a thermometer into the water.
4. After the water in the bucket boils and the thermometer has reached a stable temperature, measure the temperature of the metal sample, T_h. Stir the water in the calorimeter cup *gently* with the thermometer. When the temperature is stable, measure the temperature of the water in the calorimeter cup, T_c, which is also the temperature of the calorimeter cup.
5. *Quickly* but carefully, transfer a metal sample from the bucket to the calorimeter cup with as little splashing as possible so as not to lose any water. (Preferably transfer the sample in a paper towel so as to soak up any excess water on the sample's surface.) Stir the mixture *gently* with the thermometer and record the temperature T_e when the system reaches a stable equilibrium value.
6. Dry the metal sample and measure its mass m_s.
7. Compute the specific heat of the sample, c_s, using Eq. (4) and obtain the percent error with the accepted value.

II. LATENT HEAT OF FUSION WATER

1. Obtain the mass of the inner calorimeter cup, m_{cal}, and its specific heat c_{cal}, from part 1.
2. Fill the calorimeter cup half full of water, at a temperature 10 to 15°C above room temperature, and determine its mass. Determine the mass of the water, m_w.

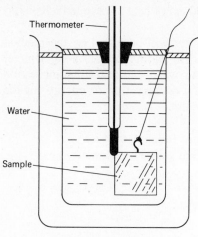

Thermometer

Water

Sample

CALORIMETER

3. Place the calorimeter cup containing the water in the outer jacket. Place a thermometer into the water, stir *gently,* and measure the temperature T_h, when it has stabilized.
4. Dry several small pieces of ice with a paper towel. *Immediately* after reading the initial temperature of the water, add the pieces of ice to the water, one at a time, without touching the piece and without splashing. *Gently* stir the ice-water mixture while slowly adding the ice; add enough ice to lower the temperature of the mixture to about 10 to 15°C below room temperature. Continue to stir gently, and record the equilibrium temperature T_e after the ice has melted completely and when the temperature of the mixture has stabilized.
5. Determine the mass of the calorimeter cup with water and melted ice from which the amount of added ice, m_i, can be determined.
6. Compute the latent heat of fusion, l_F, using Eq. (6); obtain the percent error from the accepted value.

EXPERIMENT 10

CALORIMETRY: SPECIFIC HEAT AND LATENT HEAT OF FUSION REPORT SHEET

Name _____   Section _____

Table

Part 1		Part 2	
Mass of calorimeter, m_{cal}, g		Mass of calorimeter, m_{cal}, g	
Specific heat of calorimeter, c_{cal}, cal/g·°C		Specific heat of calorimeter, c_{cal}, cal/g·°C	
Mass of calorimeter and water, g		Mass of calorimeter and water, g	
Mass of water, m_w, g		Mass of water, m_w, g	
Initial temperature of sample, T_h, °C		Initial temperature of sample, T_h, °C	
Initial temperature of water, T_c, °C		Equilibrium temperature of mixture, T_e, °C	
Equilibrium temperature of mixture, T_e, °C		Mass of calorimeter, water, and melted ice, g	
Mass of sample, m_s, g		Mass of ice, m_i, g	
Specific heat of sample, c_s		Latent heat of fusion, l_F	
c_s (exp.), cal/g·°C		l_F (exp.), cal/g	
c_s (accpt.), cal/g·°C		l_F (accpt.), cal/g	
Percent error		Percent error	

EXPERIMENT 11

ELECTROSTATICS

Materials Needed

Electroscope
Lucite or glass rod and silk cloth
Hard rubber rod and fur pelt or heavy felt cloth
Matches
Bits of paper

Background

This experiment will investigate the nature of electric charges. There are two types of electric charges: positive and negative; like charges repel each other and unlike charges attract. This lab will demonstrate the spatial dependence of electrostatic forces or electrical forces due to static electrical charges.

The principal experimental tool for this experiment is the electroscope, which consists of a circular metal plate that is in good electrical contact with a metallic vane and its holder. The vane is surrounded by a metal can but is insulated from it. (i.e., the vane is in poor electrical contact with the can). When there is no electrical charge on the electroscope, the vane should stand vertically or almost vertically without its end touching the metal support. If it does not do so, have the instructor adjust it or there will be a loss of sensitivity. Throughout the experiment make sure that the vane is able to rotate freely on the needle and that it does not rub against the support on either side. This must be regularly checked and corrected if it is wrong.

First ground the electroscope by connecting it to a grounded object. This may be done by simply touching it with your hand. The charge will leak off harmlessly through you to the earth. Between each part of this experiment, be sure to ground the electroscope so that the charge acquired from the previous part will be removed!

I. CHARGE SEPARATION

Take the lucite rod and rub it vigorously with the silk. Hold the rod horizontally about 2 or 3 cm above the plate of the electroscope and notice that the vane rotates about 40° to 50° from the vertical (on dry days). This position will henceforth be called the *extended position*. The lucite has a *positive* charge after being rubbed since it is deficient of negative electrons. The lucite attracts an excess of negative electrons in the electroscope up to the top plate, thus leaving the bottom of the electroscope with a positive charge. Since like charges repel, the vane rotates so that the positive charges can be farther away from the holder which is also positively charged. This is shown schematically in Fig. 1A. Of course, some work was done against gravity during this process since the heavy end of the vane was raised up and the light end pushed down. Where did the work to do this come from? Notice that when you move the rod away the vane returns to the vertical position (collapses) since there is no net charge on the electroscope. Now repeat this experiment with the rubber rod, which acquires a negative charge when rubbed with the fur. The pertinent schematic picture is shown in Fig. 2A.

II. DIRECT CHARGING

Next, rub the charged lucite glass rod on the plate of the electroscope to see if you can transfer some of the positive charge to the electroscope. Rerub the rod with the silk and rub the plate again to be sure it was done well. Can you get a positive charge on the electroscope by contact? Is it easy? (You can tell when there is charge left by the vane being in the extended position when the charged rod is removed.) Try next to charge the electroscope negatively by using the rubber rod the same way. Can you? Is it easy?

III. CHARGING BY INDUCTION

There is a very efficient way to get a large charge on the electroscope, known as "charging by induction." This is shown schematically in Figs. 1 and 2. Take the rubbed lucite rod and hold it 2 or 3 cm above the electroscope plate as shown in Fig. 1A. Now touch the plate with your other hand, which allows negative electrons to flow onto the plate and neutralize the positive charges on the vane and its holder. This is shown in Fig. 1B, where touching the plate is illustrated by the standard "ground" symbol. Then remove the ground (your hand) while still holding the charged rod in place, as shown in Fig. 1C. Finally, take away the rod and notice that the vane is extended, indicating a net negative charge as shown in Fig. 1D. Where did the resulting charge come from in this last step? Ground the electroscope and then repeat the process of charging by induction using the other rod with the negative charge. This procedure is illustrated in Fig. 2A, B, C, and D. Notice that when you charge the electroscope by induction, the charge left on the electroscope is opposite to the charge on the charging object.

IV. DETERMINING TYPE OF CHARGE

Suppose you are confronted with a charged electroscope and want to determine whether the charge is positive or negative. The method for doing this is shown in Fig. 3. If the electroscope has a positive charge, then the vane will collapse when the negative charged rod is brought nearby (Fig. 3A) and it will become more extended if the positively rod is brought nearby (Fig. 3B). The opposite occurs if the electroscope has a negative charge, as shown by Fig. 3C and D. Prove to yourself that this is the case by first charging the electroscope. Describe what happens in each case. One word of caution about this check is needed. If the electroscope has a small charge on it and a charged rod with opposite charge on it is brought nearby, the

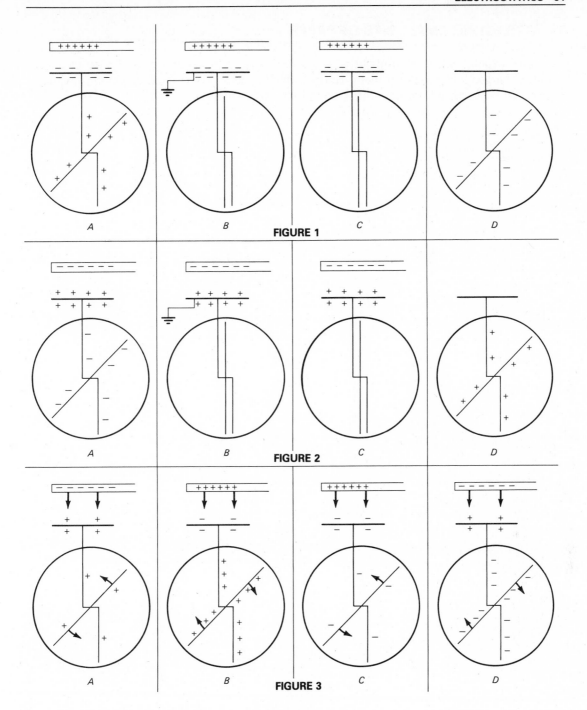

FIGURE 1

FIGURE 2

FIGURE 3

vane will first collapse and then become extended as you bring the rod nearer. Observe this process and then try to explain why it behaves this way.

V. GROUNDING BY IONIZATION

Next, charge your electroscope by induction. Then bring a lighted match near the electroscope plate. Tell what happens and try to explain why. Repeat this experiment with the electroscope charged by the other rod and again discuss it.

VI. ATTRACTING SMALL BITS OF PAPER

As a final experiment, tear off about 10 small pieces of paper about 2 or 3 mm on each side. Pass the charged lucite rod over them. Record what happens. Try to explain why this happens. Repeat the experiment with the other charged rod and explain.

EXPERIMENT 11

ELECTROSTATICS
REPORT SHEET

Name _____   Section _____

Today's procedure is broken up into six paragraphs (I through VI) of instructions. Each describes a single experiment to perform with the electroscope, and each is accompanied by a series of questions. Answer each question correctly and in detail. Your lab score will be completely determined by your answers. If you are not sure how to proceed in a particular experiment, please ask for assistance. The lucite rubbed with silk produces a positive charge on the lucite rod and the rubber rod-fur combination produces a negative charge on the rod.

Problems

I. CHARGE SEPARATION VERSUS CHARGE CONTACT

1. Explain why the vane rotates to the "extended" position even though the lucite rod is not in contact with the plate (Fig. 1A).

   _____

   _____

   Record the angle (approximately) _____

2. When the same rod is placed in contact with the plate the angle of repulsion is less. Why?

   _____

   _____

   Record the angle (approximately) _____

3. The same results occur with the rubber rod (Fig. 2A). Why?

   _____

   _____

   Record the angle _____

II. DIRECT CHARGING

1. Rubbing the lucite rod on the plate several times causes positive charge to transfer to the plate. Why?

   _____

   _____

What is the effect on the electroscope and Why?

_____

_____

Record the angle of repulsion. _____

2. What happens when the plate is rubbed with the rubber rod and why?

_____

_____

Record the angle of repulsion. _____

III. CHARGING BY INDUCTION

1. What charge flows onto the vane by the way of the ground in the presence of the lucite rod?

_____

_____

When the rod is withdrawn, what is the net charge on the electroscope?

_____

_____

Why does the vane repel? _____

Record the angle of repulsion. _____

Compare this value to the angle obtained in Sec. II. Which method of charging the plate is more efficient (faster), direct charging or charging by induction?

2. Repeat the process with the rubber rod. What is the result?

_____

_____

Record the angle of repulsion. _____

3. In both cases, how does the charge on the rod compare with the charge on the electroscope?

_____

_____

IV. DETERMINING TYPE OF CHARGE

1. Describe a method, using the electroscope, used to determine the type of charge on a vane.

   Give the method here. _____

   _____

   Explain why the method works. _____

2. Explain the second effect (end of paragraph IV) _____

V. GROUNDING BY IONIZATION

What happens when the match is brought near the plate? _____

Explain why this occurs. _____

Discussion of effect with second rod _____

_____

VI. "PAPER BITS"

1. What happens when the lucite rod is passed over the pieces of paper?

   _____

   Why does this happen? _____

   Discussion of effect with second rod _____

   _____

2. Repeat with rubber rod and explain.

   _____

   _____

EXPERIMENT 12

THE RAY BOX: REFRACTION AND REFLECTION

Materials Needed

Ray box
Optical specimen set
Centimeter ruler
Protractor
White paper and masking tape
Variac power source (12 V maximum)

Background

Under most circumstances we can analyze the behavior of light by assuming that it travels in straight lines. That is, we can use the idea of narrow light beams or "light rays." However, we must keep in mind that light is actually a wave phenomenon and that a light ray is merely a line drawn to represent the direction of advance of the wavefront of the light wave. To facilitate our understanding of refraction and reflection, we will use a special device which projects narrow beams of collimated light (i.e., "light rays") from narrow slits which we will use to explore the nature of light. This device is called a *ray box*.

When a light ray is incident on a polished surface, the ray is reflected in such a way that the angle of reflection is equal to the angle of incidence. This is called the *law of specular reflection*. This law may be derived from simple mathematical models based on the theory of electromagnetism of Maxwell. These angles are measured from a line that is perpendicular or normal to the surface. (See Fig. 2.) When light travels from one transparent media to another (such as from air to glass), the light divides, at the surface between the media, into two parts. One part of the light is reflected, and this ray obeys the law of specular reflection. The remainder of the wave energy travels through the material via the transmitted ray. If the incident ray strikes the surface of the new material with an oblique angle, the transmitted ray bends and follows a straight path along a different direction. The bending of light as it passes from one media to another is called *refraction*. This is due to the fact that the speed of light depends on the material through which the light wave passes. The speed of light c is fastest in a vacuum, being

$$c = 3.0 \times 10^8 \text{ m/s}$$

The ratio of the speed of light in a vacuum to the speed in a medium is called the *index of refraction* of the medium. This index is usually represented by the symbol *n*:

$$n = \frac{c}{v}$$

where v is the velocity of light in the medium. The index of refraction of air is very nearly 1. Crown glass has an index of refraction of 1.5. If light is incident on the surface from the medium with a larger index of refraction, the transmitted ray bends away from the perpendicular. From considerations like this or from full mathematical wave equations, we can determine the following important relationship:

$$n_1 \sin \theta_1 = n_2 \sin \theta_2$$

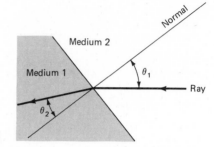

FIGURE 1

This equation is known as *Snell's law*. Again, the angles θ_1 and θ_2 shown in Figure 1 are measured from the normal or perpendicular line drawn at the surface between medium 1 and medium 2. In words, Snell's law states that the product of the index of refraction and the sine of the angle measured from the normal remains the same in each medium. Since the sine of an angle increases as the angle increases (from 0 to 90°), the angle is less in a medium whose index of refracion is greater. Thus, the light bends towrd the normal when it enters a medium with a greater index of refraction.

Procedure and Analysis of Data

1. Lay a plain white sheet of paper underneath the light box so that you get maximum coverage for the particular setup. Turn on the variac. The variac should not be set for more than 12 V.

2. Make sure the light rays do not diverge. Measure the width between the outer rays near the box and about 20 cm away. The widths should be equal. If there is a a difference, adjust the divergence control knob at the top of the box. The rays must be parallel for proper results.

3. Since we will be using the method of geometrical construction in this lab, use a sharp pencil and clean paper throughout. Reproduce ray patterns by marking tiny pencil points or "dots" along the ray paths in each experiment. Place a dot every 2 cm or so along the path. Using a ruler, draw a best-fit line through the points. It will be helpful to number each ray at various dot positions before the paper is removed. For best results the paper should be taped to the table while the construction is being made.

4. *Law of specular reflection:* Place the plane mirror in the pathway of the light rays at an oblique angle (>25°) (see Figure 2). Mark the boundary of the mirror surface and reproduce all light rays. Using a triangle, determine a normal for each light ray. Determine angle of reflection and the angle of incidence for each ray using a protractor. Estimate to a tenth of

a degree precision. Record these values and average. Have you confirmed the law of specular reflection? Determine the percentage difference between these two averaged angles.

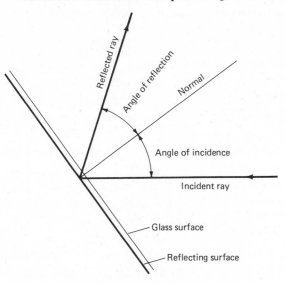

FIGURE 2

5. *Snell's law:* Air-glass-interface. Snell's law, when applied to the case of a ray of light passing from air to glass or from glass to air, may be written as

$$n_a \sin \theta_a = n_g \sin \theta_g$$

Using the definition of index of refraction for the two media, $n_a = c/v_a$ in air and $n_g = c/v_g$ in glass, the following equation may be easily derived:

$$\frac{v_a}{v_g} = \frac{\sin \theta_a}{\sin \theta_g}$$

For a discussion of refraction in terms of the different speeds of light in the air and glass, refer to Fig. 6.24 of your text and the relevant paragraphs. Aided by the above equation, in this experiment you will determine how much faster light travels in air than in glass. Place a block of glass at an oblique angle to the rays of light (see Figure 3). Mark very carefully the rays entering and leaving the block. Trace around the periphery of the block itself. Determine θ_g and θ_a as before by measuring them from the normal you construct using a triangle. Average these values and determine v_a/v_g. If $v_a = 3 \times 10^8$ m/s, what is the speed of light in glass?

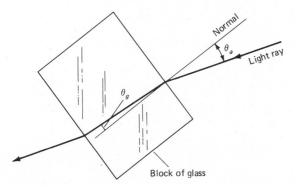

Block of glass   **FIGURE 3**

6. *Lenses and mirrors:* The focal length of a lens or mirror is defined as the point where light rays converge from a source that produces plane parallel incident light rays. The bending of light in the lens follows Snell's law, and the bending of light from the mirror's surface follows the law of specular reflection.

(a) Set up the mirror with its concave figure squarely facing the ray box (see Figure 4). The focus should be centered between the central parallel rays. Trace the paths as before. Choose one ray and draw a normal to the incident surface. Do this by first drawing a tangent to the surface at the point of incidence and then constructing a perpendicular to that line. This is the normal. By measuring the appropriate angles, show that the law of specular reflection is obeyed.

(b) Place the positive lens squarely centered in the path of the rays. The lens should be placed about 5 cm in front of the ray box. Draw the ray pattern. Mark the center of the lens. Also mark the path of the light ray inside the lens. From the drawing, determine the focal length of the lens. Choose one ray and show that Snell's law is obeyed at least qualitatively (the ray should bend toward the normal in the new medium). You will note that there is quite a bit of spherical aberration in the lens. Determine the average focal length F by averaging the central ray focal length and the outside ray focal length. What is the diameter D of the lens? From this calculate the F ratio (F/D).

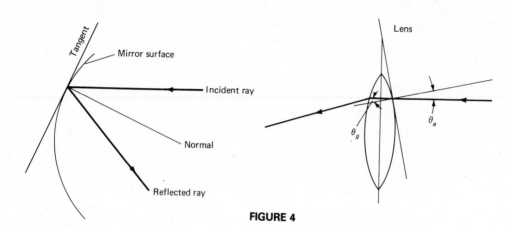

FIGURE 4

EXPERIMENT 12

THE RAY BOX:
REFRACTION AND REFLECTION
REPORT SHEET

Name _____   Section _____

I. LAW OF SPECULAR REFLECTION

	Angle of Incidence	Angle of Reflection
1.		
2.		
3.		
4.		

Average _____

% difference _____

II. SNELL'S LAW OF REFRACTION

	θ_a	θ_g
1.		
2.		
3.		
4.		
sin θ		

v_a/v_g _____

Speed of light in glass _____

III. LENSES AND MIRRORS

Central ray focal length _____ cm

Outer ray focal length _____ cm

Average focal length _____ cm

Diameter of lens _____ cm

F ratio _____ cm

Problems

1. Have you confirmed the law of specular reflection? Explain why or why not.

_____

_____

2. The speed of light is appreciably less in glass than it is in air. Explain why the light ray bends *toward* the normal as the ray passes through the air-glass interface and *away* from the normal as it leaves the glass and enters the air.

_____

_____

3. How do you know that Snell's law explains the behavior of a ray pattern in a lens? Discuss the ray as it enters and exits the lens.

_____

_____

4. (Optional.) Explain why spherical aberration occurs. How can we determine if the mirror has parabolic figure?

_____

_____

EXPERIMENT 13

SIMPLE EXPERIMENTAL TECHNIQUES

Materials Needed

25 mL 0.1 M $CoCl_2$
25 mL 0.067 M K_3PO_4
10 g crude $Ni(NH_4)_2(SO_4)_2 \cdot 6H_2O$ (containing 0.5 g $NiCO_3$)
ground to fine powder
Suction-filtration apparatus
Two plastic or glass centrifuge tubes
Disposal containers
Ice

Background

There are three common states of matter: solid, liquid, and gaseous. Since many chemical reactions and physical processes involve or produce a mixture of these states, it is useful to know how to separate them. In this experiment, you will learn common techniques for doing this, using no unusual equipment. Future experiments may call for the use of one or more of these techniques, but no explicit instructions will be given, since you will be expected to be familiar with them.

When a product is obtained in the laboratory, it is usually not pure. If it is a solid, purification is often done by *crystallization*. In this process, which is summarized in Fig. 1, the solid is soluble in a hot solvent or mixture of solvents but is insoluble (or nearly so) in the cold solvent(s). Therefore, the solid is dissolved in the hot solvent, and cooled when crystallization (formation of the solid) occurs. Any impurities present should either be very soluble in the solvent or not soluble at all. Since this is a *physical,* not chemical, process, the solvent should *not react* with the solid being crystallized.

When a purification process has been completed, it is helpful to know how efficient it is. A chemical manufacturer is not likely to use a process that gives 20 percent usable product

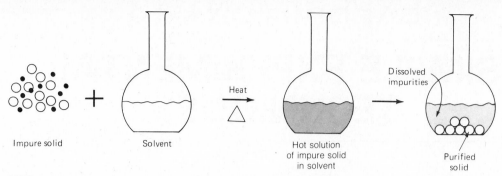

FIGURE 1   Process of crystallization. (The symbol △ stands for heat.)

and 80 percent waste unless the starting material is very cheap. A measure of efficiency in purification is percent recovery:*

$$\% \text{ recovery} = \frac{\text{yield (mass) of purified product}}{\text{theoretical yield (mass)}} \times 100\%$$

Theoretical yield represents complete recovery.

Procedure

I. GRAVITY FILTRATION

Set up the apparatus shown in Fig. 2. If necessary, ask your instructor to demonstrate the proper method of folding filter paper.

Begin heating 35 mL of H_2O in a 125-mL erlenmeyer flask (using a boiling stone). While you are waiting, weigh out 10.0 g of crude $Ni(NH_4)_2(SO_4)_2 \cdot 6H_2O$ powder. When the water boils, remove it from the heat and add the powder. Some fizzing (CO_2)* may occur. Swirl the mixture for 1 min so that most of the solid dissolves, and then pour the hot mixture into the funnel. Do not overfill. The *filtrate* that passes through the paper should be clear, leaving the solid behind on the paper. A container will be provided for collection of the solid. Allow the filtrate to cool to room temperature undisturbed, and then place it in an ice bath. The crystals that form should be suction-filtered according to directions in the next section.

II. SUCTION (VACUUM) FILTRATION

While the $Ni(HN_4)_2(SO_4)_2 \cdot 6H_2O$ is crystallizing, assemble the apparatus shown in Fig. 3. A trap should be used to prevent water from the aspirator from siphoning back into your filter flask. Have your instructor check the setup before proceeding. Fit the Buchner funnel (Fig. 3) with a circle of filter paper large enough to cover all the holes but not so large that it wrinkles or folds up on the sides of the funnel. Wet the paper with a small amount of the solvent (water) being used. Turn on the aspirator *all the way*, which should start the suction. Press down on the funnel if necessary. Transfer the purified crystals plus solvent (water) to the funnel in portions. When all the crystals have been collected, continue suctioning until they are dry. Weigh.

*When chemical reactions are involved, we calculate

$$\% \text{ yield} = \frac{\text{yield (mass) of product}}{\text{theoretical yield (mass)}} \times 100\%$$

FIGURE 2   Gravity filtration. (*Courtesy of W. K. Fife.*)

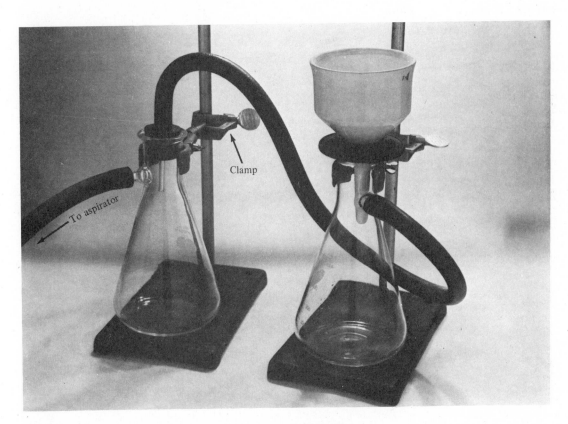

FIGURE 3   Two suction-filtration setups. (*Courtesy of W. K. Fife.*)

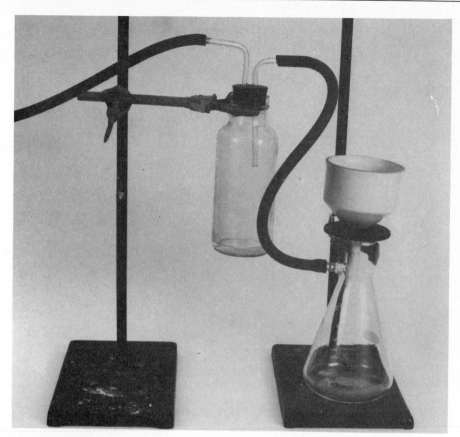

FIGURE 3   continued

FIGURE 4   Decanting: pouring off the liquid from the solid. (*Courtesy of W. K. Fife.*)

*From $NiCO_3$ impurity.

III. DECANTATION

In a 400-mL beaker, mix 25 mL of 0.1 M* $CoCl_2$ solution with 25 mL of 0.067 M K_3PO_4 solution. Then add tap water until the liquid comes up to the 300-mL mark (it need not be exact). Stir thoroughly, and pour off about 15 to 20 mL of the liquid into a 50-mL beaker for *use in the next part of the experiment*. Meanwhile, allow the remainder of the liquid to stand undisturbed until the end of the period so that the solid *precipitate* (cobalt phosphate) will settle. You should then be able to pour off (decant) the clear liquid without pouring away any of the solid, as shown in Fig. 4. A container will be provided for disposal of the solid.

IV. CENTRIFUGATION

Obtain two plastic or glass centrifuge tubes that will easily fit into the centrifuge holes. Stir the liquid you saved from part III and then pour it into the test tubes so that each is about half full. Insert the tubes into opposite holes in the centrifuge. Allow 5 min spinning time. Note the separation obtained. A container will be provided for collection of the tubes.

*A unit of concentration.

EXPERIMENT 13

SIMPLE EXPERIMENTAL TECHNIQUES
PRE-LABORATORY QUESTIONS

1. A rock that weighs 18.00 g is known to be 10 percent by weight silver. If in a separation and purification process 1.35 g of silver is recovered, what is the percent recovery?

2. Air is a mixture of O_2 and N_2 among others; suggest a way to separate these.

3. Substance A is very soluble in both hot and cold alcohol. Substance B is soluble only in hot alcohol. If you are given a sample of A contaminated with B, how could you purify it?

EXPERIMENT 13

SIMPLE EXPERIMENTAL TECHNIQUES
REPORT SHEET

Name _____   Section _____

I. FILTRATION

	Crude	Purified
1. Weighing paper $+$ $Ni(NH_4)_2(SO_4)_2 \cdot 6H_2O$	_____ . _____ g	_____ . _____ g
Weighing paper	_____ . _____ g	_____ . _____ g
$Ni(NH_4)_2(SO_4)_2 \cdot 6H_2O$	_____ . _____ g	_____ . _____ g

2. % recovery (show calculation setup)

_____ %

3. Describe the appearance of the purified product.

II. DECANTATION AND CENTRIFUGATION

1. Which of these two methods do you consider to be the most efficient method to separate a solid from a liquid? Explain.

2. What special properties must a solid have so that decantation may be used?

Problems

1. Given a mixture of salt (NaCl) and sand (SiO_2), how would you separate them?

2. A solution of three liquids, A, B, and C, is to be distilled. Upon distillation, the liquid with the lowest boiling point will boil first, at its boiling point. B boils 10°C above A, and C boils 15°C above B. On distillation, liquid begins to distill at 80°C. Indicate the order—and the temperatures—at which the three liquids distill. Explain your reasoning.

EXPERIMENT 14

ELEMENTS

Materials Needed

Large cardboard or posterboard periodic tables which are identified *only* by names of the elements; the space for each element should be sufficiently large that a glass vial containing a sample may be placed there
Vials of elements, identified *only* by the symbol
0.02 M $CuSO_4$ solution
"Household ammonia" (approximately 3 M)
Small vial containing several grams of baking soda
Vinegar (5% acetic acid, w/v, in H_2O)
Balloons which will fit over a 250 flask fitted with a one-hole rubber stopper

Background

The concept that all matter is composed of a few fundamental substances began in ancient Greece with Empedocles, who in 440 B.C. proposed that all substances were some combination of earth, air, fire, and water. There was no experimental evidence for his idea; nevertheless, this concept of four simple "elements" persisted for centuries.* Before the Middle Ages, only nine true elements had been discovered (none recognized as such), and many more false ones, including water and salt, were presumed to be elements. Elements that are commonplace now were completely unknown:†

> In the Roman's luxurious display of alabaster floors, marble stairs and mosaic ceilings, no nickel-plated or chromium fixtures were to be seen, and though he might have golden bowls, he could not buy even the smallest aluminum trinket. There were no lanterns to walk along the splendid lava pavements of the city streets at night, for the white glow of the tungsten filament and crimson glow of the neon tube were lacking. The water that came to him through miles of magnificent aqueducts was a menace to health, for there was no chlorine with which to kill the bacteria, and if he lay grasping for breath, no cylinder of oxygen to save him.

*A modern definition: "An element is a pure substance which cannot be broken down into simpler substances by chemical means."
†Mary E. Weeks, *Discovery of the Elements,* 7th edition, 1968.

Many of the real elements known to Greeks and Romans were those which occur uncombined in nature, such as carbon, sulfur, silver, and gold. Salt is not an element, but the means of breaking down this *compound* to sodium and chlorine was not discovered until the 19th century. Sand is among the most common substances known; it is mainly a compound of silicon and oxygen, but no one succeeded in isolating the silicon from it until 1824. In fact, most elements occur in nature as compounds—combined with other elements; there are no naturally occurring chunks of elemental sodium, calcium,* aluminum, iodine, or zinc to be found.

For many centuries, there was no uniform way of representing elements, and alchemical symbols such as ☿ = mercury, ☾ = silver, and ♄ = lead were used. Not until 1814 were simple letters proposed (by Berzelius) to represent the elements, many of which had names originating in Latin, e.g., Fe = ferrum = iron, and Pb = plumbum = lead.

At about the time Empedocles proposed the four fundamental substances, Democritus proposed the existence of *atoms*: extremely tiny particles which could not be further divided.† Again, there was no experimental evidence for atoms, and the idea was not widely accepted. An accurate picture of the nature of atoms was even slower to develop than that of elements, since atoms were too small to be "seen" and therefore more abstract. It was not until the end of the 19th and beginning of the 20th century that experiments by physicists demonstrated the nature of atoms.

In the 1700s, accurate, quantitative measurements of mass by Lavoisier showed that matter is neither lost nor gained during a chemical change: *the law of conservation of matter*. Why is this true? It is now known that atoms cannot be created or destroyed in chemical changes. Since atoms are conserved in chemical changes, and matter is composed of atoms, mass does not change during a chemical reaction.

Procedure

I. ELEMENTS: SYMBOLS AND PHYSICAL PROPERTIES

At your laboratory station you will find a chart, called a *periodic table,* listing the names of elements, and arranged in a certain order. In the glass vials are samples of pure elements, each with its symbol printed on the vial. Do *not* remove any element from its vial until instructed to do so.

By matching the symbol on the vial with the name on the periodic table, place each element in its proper location. Record a description of each element.

II. ELEMENTS: DENSITY

Some elements are gases; others are dense solids (few are liquid at room temperature—can you think of any?). Which element on your chart do you think is most dense? Find out, by measuring the density of at least two elements of your choice. If necessary, review Density in Experiment 1 on Measurement.

Caution: Check with the instructor before proceeding to ensure that you have not chosen a reactive element.

Find the mass of your element sample by weighing a plastic weighing boat, adding the element (be sure to use the entire vial), and reweighing. Find the volume of your element by approximately half-filling a 25-mL graduated cylinder with distilled water, reading the volume

---

*Teeth and bones contain *compounds* of calcium, *not* elemental calcium.

†A modern definition: "An atom is the smallest particle of an element which retains the properties of that element."

FIGURE 1 Flask with vial. (*Courtesy of J. Douglas Bartlow.*)

to the nearest 0.1 mL, and then adding the element sample. Read the new volume, and find the volume of your element by difference. Calculate the density.

Repeat the above procedure with a different element. When you are finished, dry off both samples on a paper towel.

III. CONSERVATION OF MATTER

1. Using a 50-mL graduated cylinder, pour 50 mL 0.02 M $CuSO_4$ (copper sulfate) into a 250-mL flask. Separately, pour 25 mL "household ammonia" (aqueous NH_3) into a 25-mL graduated cylinder. Weigh the flask with $CuSO_4$ solution and then place the graduated cylinder of ammonia on the platform balance and weigh. You will need to add these masses together on the report sheet. Now, pour the ammonia into the $CuSO_4$ solution, being careful not to spill any. What evidence is there of a chemical reaction? Weigh the flask and now-empty cylinder.

2. Wash out the 250-mL flask from step 1 with distilled water, and dry the outside. Pour in 50 mL of vinegar, and place the flask with vinegar on the balance. Place a vial of baking soda on the balance as well. Weigh. Now, carefully lower the vial (tweezers may be helpful) into the flask (Fig. 1). Swirl to mix, making sure the vinegar gets into the vial. Allow the contents to stand, occasionally swirling, until all reaction appears to have stopped and no solid remains in the vial. Reweigh. Rinse out the flask and vial with distilled water when finished.

3. Repeat the experiment in step 2 except *immediately* after lowering the vial into the flask, attach a *weighed* one-hold rubber stopper fitted to a balloon to the top of the flask.

EXPERIMENT 14

ELEMENTS
PRE-LABORATORY QUESTIONS

1. The method of finding density in this experiment cannot be used for all elements. Explain the limitations on this method.

2. Mercury is a very dense element (13.6 g/mL), but we did not determine it experimentally because of its toxicity. Calculate the volume occupied by one pound (453.6 g) of mercury. Note that 1 lb of water occupies 453.6 mL, or about 1 pint.

EXPERIMENT 14

ELEMENTS
REPORT SHEET

Name_____   Section_____

I. ELEMENTS: SYMBOLS AND PHYSICAL PROPERTIES

In the space below, record the symbol, name, and a brief description of each sample. Classify the elements according to position on the periodic table, e.g., alkali metal, transition metal, halogen. Note that some elements do not have "classifications."

II. ELEMENTS: DENSITY

Element 1 _____

Mass of sample + weighing paper	_____ g
Mass of empty weighing paper	_____ g
Mass of element sample	_____ g
Volume of water + sample	_____ mL
Initial volume of water	_____ mL
Volume of element sample	_____ mL
Density	_____ g/mL

Show calculation setup

Element 2 _____

Mass of sample + weighing paper	_____ g
Mass of empty weighing paper	_____ g
Mass of element sample	_____ g
Volume of water + sample	_____ mL
Initial volume of water	_____ mL
Volume of element sample	_____ mL
Density	_____ g/mL

Again, shown calculation setup

III. CONSERVATION OF MATTER

1. Reaction of $CuSO_4$ with ammonia: What evidence is there of a reaction?

   (a) Mass of reactants

   (i) Mass of 25-mL graduated cylinder with NH_3 _____ g

   (ii) Mass of 250-mL flask with $CuSO_4$ _____ g

   (iii) Total mass of reactants _____ g

(b) Mass of products

    (i) Mass of empty 25-mL graduated cylinder   _____ g

    (ii) Mass of 250-mL flask with mixture   _____ g

    (iii) Total mass of products   _____ g

(c) Is the above total masses what you expected? Explain.

2. Reaction of baking soda with vinegar (no balloon): What evidence is there of a reaction?

    (a) Mass of products + containers after reaction   _____ g

    (b) Mass of reactants + containers before reaction   _____ g

    (c) Is the above data what you expected? Explain.

3. Reaction of baking soda with vinegar (with attached balloon)

    (a) Mass of products + containers after reaction   _____ g

    (b) Mass of reactants + containers before reaction   _____ g

    (c) Does the above data differ from the result in question 2? Is this expected? Explain.

Problems

1. In examining your descriptions of the element samples, what general trends do you notice in the physical properties of metals versus nonmetals?

2. In a given reaction, if the total mass of reactants weighs 3.6 g and one product weighs 2.2 g, what does the other product weigh?

EXPERIMENT 15

OBSERVING MATTER

Materials Needed

Unknowns
50-mL solutions of each, labeled A to G
A. 0.5 M Bi(NO$_3$)$_3$ in 1 M HNO$_3$
B. 0.1 M Cu(NO$_3$)$_2$
C. 0.2 M Na$_2$CO$_3$
D. 6 M HNO$_3$

E. 6 M NH$_3$
F. 1 M NaI
G. distilled water

20 mL of each, labeled 1 to 4
1. Vegetable oil and water
2. Distilled water
3. Dirt and water
4. Salt solution

10 mL of each, labeled as follows:
0.5 M Bi(NO$_3$)$_3$ in 1 M HNO$_3$
1.0 M NaI

Other
10 mL vegetable oil
3 g salt
1 small piece of Li
1 chip of dry ice
10 cm Cu wire
95% ethanol

2 g tin shot
1 chip of charcoal
1 g sugar
9-volt battery with Cu wire leads
10 mL fresh, 6% H$_2$O$_2$

Background

If you are ill, you go see the doctor.

If a crime is committed, investigators come to the scene.

Chemists use transparent test tubes.

To assess the extent of a natural disaster, the governor visits the site.

All of these statements have one thing in common: *observation*.

Foremost in the study of chemistry is observation. The chemist's oldest and still most important tools for observation are the human senses. Laboratory instruments are simply tools that help our senses detect changes matter undergoes.

The purpose of this experiment is to train you to observe carefully using your eyes, touch, nose, ears. (Taste used to be required, but too many chemists gave their lives to satisfy this requirement; so, do not taste anything in the laboratory—not even that soft drink; keep it out!)

Procedure

I. SEVEN TEST TUBES

You will be given seven test tubes each containing a different chemical. You are to design some orderly fashion of mixing two at a time and observe the results. Here is what you should do.

1. Decide on a scheme to help you keep track of the mixing process. Make sure to test all possible combinations.
2. Label the samples you are given from 1 through 7, or A through G.
3. Obtain small clean test tubes and arrange them in front of you in a rack.
3. Following your scheme, mix two samples at a time taking an eye dropper full of each sample.
5. Carefully observe for any indication of a chemical reaction: heat changes, color changes, gas evolution, odor given off, precipitation, fizzing sound.
6. Record your observations on your scheme. Identify all test tubes containing active ingredients and the one which does not react at all.

II. A SURPRISE

Take a spatula full of solid NH_4NO_3, ammonium nitrate, and put into a normal-sized test tube. Add about as much water as there is solid in the test tube—no more! Shake vigorously and feel the bottom of the tube.

III. PURE SUBSTANCES VERSUS MIXTURES

You will be given four samples, each of which you should identify as a pure substance, a homogeneous mixture, or a heterogeneous mixture. You can do this by simply looking at the sample, or by applying heat to it too. If you cannot decide visually whether it is a pure substance or a mixture, take a small amount, perhaps 10 to 20 mL of it, place in a small beaker (about 100 mL), and heat *almost* to dryness. Remove the hot plate, allow to cool, and observe. In the following table write the observations and the reasons for your identification of a given sample as pure substance or mixture.

Sample	Pure Substance?	Mixture? Homogeneous	Mixture? Heterogeneous	Reason
1				
2				
3				
4				

IV. PHYSICAL AND CHEMICAL CHANGES

How do we know if a change matter undergoes is physical or chemical? Physical changes do not alter the basic chemical composition. These may be changes in state, shape, or size. A chemical change results in a totally new substance. "New substance" means it now has a different chemical makeup.

We can sometimes tell whether a change is physical or chemical by just observing what is happening. Melting ice or breaking a stick simply changes the state or size of the material; and so does grinding wheat into flour. After each change, the substance is still water, wood, and wheat. If direction observation is not enough, we may have to do a little test.

What if wood rots, or a wheat storage elevator explodes? These are chemical changes that produce something other than wood and wheat.

In the table below you are given some assignments to observe changes. Under *Process* you are given instructions for three substances to be mixed with water and four substances to be heated. Under *Change* note whether it is a physical change, a chemical change, or no change at all. Under *Reason* indicate what led to your conclusion. This may be simple observing, an evaporation, a filtration, common sense. (Use the techniques learned in Experiment 13).

Process	Change	Reason
Oil + water Pour about 10 mL of vegetable oil into about 10 mL of water.		
Salt + water Take a spatula full of salt and mix into about 10 mL of water in a small beaker. Stir well.		
Li + water With forceps remove a small piece of Li, clean off any oil, and drop into a small beaker containing 10 mL water.		

(continued)

Process	Change	Reason
Dry ice + heat DO NOT TOUCH THE DRY ICE! Using gloves or tongs put a small piece of dry ice on the countertop and observe.		
Cu + heat Take about 10 cm of Cu wire, hold it with tongs on one end, and put the other end into a strong flame. When glowing hot, take out the flame, and watch its color change.		
Peroxide + heat Pour 5 to 10 mL of peroxide, H_2O_2, into a small beaker. Place on a ring stand, *warm gently*, and observe.		
Ice + heat Put a chip of ice into a small beaker and heat.		

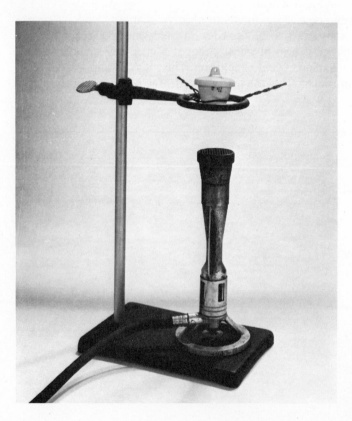

FIGURE 1   Heating in a crucible.
(*Courtesy of W. K. Fife.*)

V. ELEMENT OR COMPOUND

You will be given four samples. They are water, tin, sugar, and coal. Using two sets, the effect of heat and the effect of electricity, you can deduce whether the material is a compound or an element. If heat or electricity produces no change (NC), or only a change of state (CS), the material must be an element. If a new substance (NS) is produced as evidenced by a color change, odor given off, effervescence, etc., you may assume that a compound is being decomposed.

To test the effect of heat, place the tin, sugar, and coal into separate crucibles and heat slowly and gently (see Fig. 1). You need not test the effect of heat on water (presumably you know what happens). Set aside any sample which showed chemical change.

To test the effect of electricity on the remaining samples proceed as follows. First, test each sample for conductivity using an ohmmeter. Set aside any sample which shows conductivity. Test the remaining samples for electrical decomposition by placing the wire leads from a 9-V battery into the sample. *Caution: Do not allow the leads to touch each other!*

	Effect of		
	Heat	Electricity	
Sample			Conclusion
Tin			
Sugar			
Coal			
Water			

VI. COMPOSITION (Start this section early since filtration takes time)

Four of you will each be assigned one of the samples from the table below. You will all share the results. The amounts given in the instructions below are for student 1.

Into a large test tube pour 2.0 mL of the solution labeled "+" and 2.0 mL of the solution labeled "−." You may want to use an eye dropper to adjust the volume to the exact level. Filter the precipitate using suction filtration and the preweighted filter paper. Pour two 5-mL portions of ethanol over the precipitate and allow the suction filtration to "dry" the precipitate. Then dry in an oven or with a very low flame. When dry, weigh the filter paper with the sample. Enter the data below and share the results with the others in your group.

			Weights of Dry		
Student	Solution "+"	Solution "−"	Ppt. + Filt. Paper	− Filt. Paper	= Ppt.
1	2.0 mL	+ 2.0 mL			
2	3.0 mL	+ 2.0 mL			
3	4.0 mL	+ 2.0 mL			
4	0.5 mL	+ 2.0 mL			

What conclusion can you draw?

EXPERIMENT 15

OBSERVING MATTER
PRE-LABORATORY QUESTIONS

1. What method of observation should never be used in the laboratory?

2. From the procedure, list at least four indications of an observed reaction.

3. Explain the difference between a homogeneous and a heterogeneous mixture.

EXPERIMENT 15

OBSERVING MATTER
REPORT SHEET

Name _____   Section _____

I. SEVEN TEST TUBES

Record your observations on your scheme.

Identify the test tube which does not react _____.

II. A SURPRISE

III. PURE SUBSTANCES VERSUS MIXTURES

Sample	Pure Substance?	Mixture		Reason
		Homogeneous	Heterogeneous	
1				
2				
3				
4				

IV. PHYSICAL AND CHEMICAL CHANGES

Process	Change	Reason
Oil + water Pour about 10 mL of vegetable oil into about 10 mL of water.		
Salt + water Take a spatula full of salt and mix into about 10 mL of water in a small beaker. Stir well.		
Li + water With forceps remove a small piece of Li, clean off any oil, and drop into a small beaker containing 10 mL water.		

133

(continued)

Process	Change	Reason
Dry ice + heat DO NOT TOUCH THE DRY ICE! Using gloves or tongs put a small piece of dry ice on the countertop and observe.		
Cu + heat Take about 10 cm of Cu wire, hold it with tongs on one end, and put the other end into a strong flame. When glowing hot, take out the flame, and watch its color change.		
Peroxide + heat Pour 5 to 10 mL of peroxide, H_2O_2, into a small beaker. Place on a ring stand, *warm gently,* and observe.		
Ice + heat Put a chip of ice into a small beaker and heat.		

V. ELEMENT OR COMPOUND

Sample	Effect of		Conclusion
	Heat	Electricity	
Tin			
Sugar			
Coal			
Water			

VI. COMPOSITION

Student	Solution "+"		Solution "−"	Weights of Dry		
				Ppt. + Filt. Paper	− Filt. Paper	= Ppt.
1	2.0 mL	+	2.0 mL			
2	3.0 mL	+	2.0 mL			
3	4.0 mL	+	2.0 mL			
4	0.5 mL	+	2.0 mL			

What conclusion can your draw?

EXPERIMENT 16

CHEMICAL REACTIONS

Materials Needed

Experiment 1: 4 mL 0.1 M CuSO$_4$, 7 mL 0.1 M KOH
Experiment 2: Small marble chip (0.5 g or less), 6 mL 6 M HCl
Experiment 3: Small lump of Ca, 6 mL distilled H$_2$O
Experiment 4: Small piece of Mn, 8 mL 1.0 M HClO$_4$
Experiment 5: 3 mL 0.2 M H$_2$O$_2$,* 5 mL 0.2 M Ce(ClO$_4$)$_4$*
Experiment 6: 7 mL 0.05 M KCr(OH)$_4$,* 5 mL 0.1 M H$_2$SO$_4$
Experiment 7: 3 mL 0.03 M MnSO$_4$, 3 mL 0.02 M KMnO$_4$
Experiment 8: 6 mL 0.2 M Ce(ClO$_4$)$_4$,* 3 mL 0.2 M H$_2$C$_2$O$_4$
Experiment 9: 4 mL 0.2 M H$_2$O$_2$,* 1 mL 0.5 M KOH, 4 mL 0.05 M KCr(OH)$_4$*
Experiment 10: 4 mL 0.1 M CuSO$_4$, 3 mL 0.1 M SO$_2$, 4 mL 0.1 M HI

Background

An important feature of chemistry, as well as other sciences, is the value of accurate observations. It is by observation that the properties of all substances are known. Chemical and physical changes in these substances can often be followed by careful observation. For example, if the interaction of A and B produces a pale yellow-green gas, and chlorine is known to have these properties, then this interaction of A and B might be producing chlorine as a product.

Properties of known substances are listed in tables, usually in reference books such as the *Handbook of Chemistry and Physics*. A partial listing is provided in Table 16-1 at the end of this experiment.

You will perform several experiments involving chemical changes (reactions), and you must carefully observe the results. On the basis of what you see, you should be able to identify

*Special solutions:
0.2 M Ce(ClO$_4$)$_4$: Dissolve 109.6 g of (NH$_4$)$_2$Ce(NO$_3$)$_6$ in a mixture of 600 mL of H$_2$O and 150 mL of 71% HClO$_4$. Dilute to 1 liter with H$_2$O. The solution is somewhat light-sensitive and should be stored in dark bottles.
0.05 M KCr(OH)$_4$: Dissolve 20.0 g of Cr(NO$_3$)$_3$ · 9 H$_2$O in H$_2$O. Separately, dissolve 20.0 g of KOH in H$_2$O in a 1-liter volumetric flask. Add the Cr(NO$_3$)$_3$ solution to the KOH solution with stirring, when all material should dissolve to a deep green solution. Dilute to 1 liter with H$_2$O. The solution cannot be stored.
0.2 M H$_2$O$_2$: Dissolve 20 mL of 30% H$_2$O$_2$ in H$_2$O. Dilute to 1 liter with H$_2$O.

Table 16-1 Properties of some compounds and elements

Symbol	Name	Properties
Ca	Calcium (metal)	Silvery metal. Insoluble in H_2O but reacts with it, giving H_2.
$CaCl_2$	Calcium chloride	Soluble in H_2O, giving a colorless solution. *Test* with $AgNO_3$: gives a precipitate.
$Ca(OH)_2$	Calcium hydroxide	White solid. Insolube in H_2O but in contact with it will cause phenolphthalein to turn pink (*test*).
$Ca(NO_3)_2$	Calcium nitrate	Soluble in H_2O, giving a colorless solution. *Test* with $AgNO_3$: does not give a precipitate.
Ca_3N_2	Calcium nitride	Brown solid. Insoluble in H_2O but reacts violently with it.
$Ca(ClO_4)_2$	Calcium perchlorate	Soluble in H_2O, giving a colorless solution. *Test* with $AgNO_3$: does not give a precipitate.
C	Carbon	Black solid, insoluble in H_2O.
CCl_4	Carbon tetrachloride	Colorless liquid, insoluble in and denser than H_2O.
CO_2	Carbon dioxide	Colorless, odorless gas, slightly soluble in H_2O. Does not burn, and extinguishes a glowing splint.
Ce	Cerium (metal)	Gray, lustrous metal. Insoluble in H_2O but slowly reacts, releasing H_2.
$CeCl_3$	Cerium(III) chloride	Soluble in H_2O, giving a colorless solution. *Test* with $AgNO_3$: gives a precipitate.
$CeCl_4$	Cerium(IV) chloride	Soluble in H_2O, giving an orange solution. *Test* with $AgNO_3$: gives a precipitate.
$Ce(OH)_3$	Cerium(III) hydroxide	White solid, insoluble in H_2O.
$Ce(OH)_4$	Cerium(IV) hydroxide	Yellow solid, insoluble in H_2O.
$Ce(ClO_4)_3$	Cerium(III) perchlorate	Soluble in H_2O, giving a colorless solution. *Test* with $AgNO_3$: does not give a precipitate.
Cl_2	Chlorine	Yellow-green gas with a pungent odor. Slightly soluble in H_2O.
HCl	Hydrogen chloride	Soluble in H_2O, giving a colorless solution known as hydrochloric acid. *Test* with $AgNO_3$: gives a precipitate. Turns Congo Red paper blue (*test*).
$HClO_4$	Perchloric acid	Soluble in H_3O, giving a colorless solution. *Test* with $AgNO_3$: does not give a precipitate. Turns Congo Red paper blue (*test*).
Cr	Chromium (metal)	Bright, lustrous metal, insoluble in H_2O.
$CrCl_2$	Chromium(II) chloride	Soluble in H_2O, giving a pale blue solution. *Test* with $AgNO_3$: gives a precipitate.

Table 16-1 (*continued*)

Symbol	Name	Properties
$CrCl_3$	Chromium(III) chloride	Soluble in H_2O, giving a dark green solution. *Test* with $AgNO_3$: gives a precipitate.
$Cr(OH)_2$	Chromium(II) hydroxide	White solid, insoluble in H_2O.
$Cr(OH)_3$	Chromium(III) hydroxide	Dark green or dark greenish-blue solid. Insoluble in H_2O.
CrO_3	Chromium trioxide	Soluble in H_2O, giving an orange solution. *Test* with $BaCl_2$: gives *no* precipitate. Turns Congo Red paper blue (*test*).
$CrSO_4$	Chromium(II) sulfate	Soluble in H_2O, giving a pale blue solution. *Test* with $BaCl_2$: gives a precipitate.
$Cr_2(SO_4)_3$	Chromium(II) sulfate	Soluble in H_2O, giving a dark green or dark greenish-blue solution. *Test* with $BaCl_2$: gives a precipitate.
K_2CrO_4	Potassium chromate	Soluble in H_2O, giving a yellow solution. *Test* with $BaCl_2$: gives a precipitate. Does *not* turn Congo Red paper blue (*test*).
Cu	Copper (metal)	Reddish, lustrous metal. Insoluble in H_2O.
$CuCl_2$	Copper(II) chloride	Soluble in H_2O, giving a blue solution. *Test* with $AgNO_3$: gives a precipitate.
$Cu(OH)_2$	Copper(II) hydroxide	Blue solid, insoluble in H_2O.
CuI	Copper(I) iodide	Off-white or beige solid, insoluble in H_2O.
$Cu(NO_3)_2$	Copper(II) nitrate	Soluble in H_2O, giving a blue solution. Does not give a precipitate with either $AgNO_3$ or $BaCl_2$ (*test*).
Cu_2O	Copper(I) oxide	Reddish-orange solid, insoluble in H_2O.
CuO	Copper(II) oxide	Black solid, insoluble in H_2O.
$CuSO_4$	Copper(II) sulfate	Soluble in H_2O, giving a blue solution. *Test* with $BaCl_2$: gives a precipitate.
CuS	Copper(II) sulfide	Black solid, insoluble in H_2O.
H_2	Hydrogen	Colorless, odorless gas, insoluble in H_2O. Burns or pops when ignited with a burning splint.
HCl	Hydrogen chloride	Soluble in H_2O, giving a colorless solution known as hydrochloric acid. *Test* with $AgNO_3$: gives a precipitate. Turns Congo Red paper blue (*test*).
$HClO_4$	Perchloric acid	Soluble in H_2O, giving a colorless solution. *Test* with $AgNO_3$: does not give a precipitate. Turns Congo Red paper blue (*test*).
$HMnO_4$	Permanganic acid	Soluble in H_2O, giving a very dark purple solution.
HNO_3	Nitric acid	Soluble in H_2O, giving a colorless solution. Does not give a precipitate with either $AgNO_3$ or $BaCl_2$ (*test*). Turns Congo Red paper blue (*test*).

(continued)

Table 16-1   Properties of some compounds and elements (*continued*)

Symbol	Name	Properties
H_2SO_4	Sulfuric acid	Soluble in H_2O, giving a colorless solution. *Test* with $BaCl_2$: gives a precipitate. Turns Congo Red paper blue (*test*).
I_2	Iodine	Black or dark violet solid, insoluble in H_2O.
Mn	Manganese (metal)	Lustrous, gray metal, insoluble in H_2O.
$HMnO_4$	Permanganic acid	Soluble in H_2O, giving a very dark purple solution.
$MnCl_2$	Manganese(II) chloride	Soluble in H_2O, giving a colorless or very pale pink solution. *Test* with $AgNO_3$: gives a precipitate.
$Mn(OH)_2$	Manganese(II) hydroxide	White solid, turning brown on standing. Insoluble in H_2O.
$Mn(NO_3)_2$	Manganese(II) nitrate	Soluble in H_2O, giving a colorless solution. Does not give a precipitate with either $AgNO_3$ or $BaCl_2$.
$Mn(ClO_4)_2$	Manganese(II) perchlorate	Soluble in H_2O, giving a colorless solution. Does not give a precipitate with either $AgNO_3$ or $BaCl_2$.
MnO_2	Manganese dioxide	Black or dark brown solid, insoluble in H_2O.
MnS	Manganese(II) sulfide	Beige solid, insoluble in H_2O.
N_2	Nitrogen	Colorless, odorless gas. Does not burn and will extinguish a glowing splint.
NO_2	Nitrogen dioxide	Dark brown gas, insoluble in H_2O.
HNO_3	Nitric acid	Soluble in H_2O, giving a colorless solution. Does not give a precipitate with either $BaCl_2$ or $AgNO_3$ (*test*). Turns Congo Red paper blue (*test*).
NH_3	Ammonia	Soluble in H_2O, giving a colorless solution, with a characteristic odor if concentrated enough. Turns pink with phenolphthalein (*test*). Turns purple with $CuSO_4$ (*test*).
O_2	Oxygen	Colorless, odorless gas. Ignites a glowing splint.
K	Potassium (metal)	Silvery metal. Reacts extremely violently with H_2O or aqueous solutions.
KCl	Potassium chloride	Soluble in H_2O, giving a colorless solution. *Test* with $AgNO_3$: gives a precipitate.
K_2CrO_4	Potassium chromate	Soluble in H_2O, giving a yellow solution. *Test* with $BaCl_2$: gives a precipitate. Does not turn Congo Red paper blue (*test*).
KOH	Potassium hydroxide	Soluble in H_2O, giving a colorless solution. Turns pink with phenolphthalein (*test*). Does *not* turn deep purple with $CuSO_4$ (*test*).

Table 16-1   (continued)

Symbol	Name	Properties
KNO_3	Potassium nitrate	Soluble in H_2O, giving a colorless solution. Does *not* give a precipitate with either $BaCl_2$ or $AgNO_3$ (*test*).
K_2SO_4	Potassium sulfate	Soluble in H_2O, giving a colorless solution. *Test* with $BaCl_2$: gives a precipitate.
K_2S	Potassium sulfide	Soluble in H_2O, giving a colorless solution. *Test* with $AgNO_3$: gives a *black* precipitate.
S	Sulfur	Pale yellow solid, insoluble in H_2O.
H_2S	Hydrogen sulfide	Colorless gas, with an offensive, rotten-egg odor. Slightly soluble in H_2O.
H_2SO_4	Sulfuric acid	Soluble in H_2O, giving a colorless solution. *Test* with $BaCl_2$: gives a precipitate. Turns Congo Red paper blue (*test*).

the products of the reaction. It is not necessary, or in this case even desirable, that you be able to predict the products of the reaction before doing the experiment. All that is necessary is that you observe carefully. Finally , you will write a chemical equation, based on what you have observed.

Example 1.   Suppose you did the following experiment. (But do not actually do it.) You start with CuO [copper(II) oxide] and an aqueous solution of H_2SO_4 (sulfuric acid). These two substances are *reactants*—things present initially that undergo chemical change to form products (which are not identical with the reactants).

The CuO, a black, powdery solid, is placed in a test tube. The clear, colorless H_2SO_4 solution is added. On stirring or shaking, the black solid reacts, disappearing into the solution. A clear blue liquid results. No gases are evolved, and no precipitate formed. What has happened in the reaction?

You know the reactants were CuO and H_2SO_4, and reactants are placed on the left-hand side of the arrow in a chemical equation:

$$CuO + H_2SO_4 \rightarrow ??$$

Since matter cannot be created or destroyed in chemical reactions (Law of Conservation of Mass), *all elements present originally* (left-hand side of equation) *must also appear in the products* (right-hand side of equation). First, consider the Cu atoms. Where did they go? Consult the table of properties for Cu compounds. You will find that $CuCl_2$ forms a blue solution, such as that stated in the example. But this compound cannot possibly be present since none of the reactants contained any Cl atoms. All the necessary atoms for the formation of $Cu(OH)_2$ are present, but this is a blue *precipitate*, not a solution. Only $CuSO_4$ fits the description, so it must be one of the products.

You would confirm the presence of $CuSO_4$ by testing with $BaCl_2$ solution, as suggested in the Table of Properties. Directions for doing this are given in part III, Special Tests.

What about H atoms? They did not form any H_2, since no gas was formed. The other product is water, H_2O. Since the presence of water as a product (or reactant) is not easily deduced, you will be told if water participates as a reactant or is a product in any of the experiments that you do.

You now have

$$CuO + H_2SO_4 \rightarrow CuSO_4 + H_2O$$

In this case, the equation is balanced as written.

Example 2. Aqueous solutions of N_2H_4 (hydrazine) and CrO_3 (chromium trioxide) are mixed. A colorless, odorless gas is evolved. It does not burn and does not ignite a glowing splint. In addition, the remaining liquid becomes cloudy green, and solid particles can be seen. The mixture is centrifuged to separate the solid, which is seen to be dark green. A clear, colorless *supernatant* liquid remains above the precipitate. The partial equation therefore is:

$$N_2H_4 + CrO_3 \rightarrow \text{??}$$

First, the gas should be investigated. What was it? If it had been H_2, it would have burned when ignited. If it had been O_2, it would have ignited a glowing splint. No NO_2 was present because the gas was colorless, not dark brown. H_2S, Cl_2, and CO_2 may all be excluded as possible gases, since the reactants did not contain any atoms of S, Cl, or C. What gas remains? N_2 is the only possibility. You now have

$$N_2H_4 + CrO_3 \rightarrow N_2 + \text{??}$$

To do the previous kind of deduction, you should ask yourself, "What *gases* are possible from some combination of atoms of N, H, Cr, and O?"

Now, consider the precipitate. What is dark green, insoluble, and could have been formed from only N, H, Cr, and O? The only possibility is $Cr(OH)_3$. Consult Table 16-1 to convince yourself that this is true.

$$N_2H_4 + CrO_3 \rightarrow N_2 + Cr(OH)_3 + \text{??}$$

What else may have been formed? Examine the clear, colorless solution (supernatant liquid above the precipitate). What has this color, is a solution (soluble in H_2O), and could come from N, H, Cr, and O? From the chart, you should be able to see that the only possibilities (other than H_2O) are HNO_3 and NH_3. HNO_3 requires a test with Congo Red paper (see Table 16-1 and also part III, Special Tests). On doing this test, you would find a negative result (no blue color) and would conclude that HNO_3 cannot be a product. Likewise, the phenolphthalein test indicated for the NH_3 would also be negative.

Since you have no evidence for any other products, you should therefore conclude that no other products were formed. In addition, notice that all types of atoms (N, H, Cr, and O) have been accounted for. All that remains is to balance the equation (this may require practice):

$$3 N_2H_4 + 4 CrO_3 \rightarrow 3 N_2 + 4 Cr(OH)_3$$

Procedure

I. GENERAL (READ CAREFULLY!!)

It is very important that the directions given in each specific experiment be followed carefully and *explicitly*. Otherwise, unusual products may arise. Observe and record all occurrences.

1. *Gas formation.* If a gas is formed, *cautiously* note its odor, if any. Have a splint ready to test for H_2 or O_2 (see part III, Special Tests). If a gas is formed, wait until its evolution

is finished before making any other tests. A few tiny bubbles do *not* constitute a gas and may be ignored.

2. *Precipitate formation*. If the reaction mixture looks cloudy, or you can see solid particles after all reaction has stopped, centrifuge the mixture (be sure to balance the tubes). Observe the solid product. What color is it?

3. *The solution*. Ask yourself, "What compound(s) could be present in a solution of this color and contain(s) only those atoms present in the reactants?" Consult the table of properties (Table 16-1) for all possibilities (make a list if necessary). *Eliminate as many choices as possible on the basis of color*. For those possible substances that remain, perform any special tests suggested in the table of properties. A substance you think is present must fulfill *all* special tests in order to be confirmed as a product. However, **do not** simply add *every available chemical listed in the Special Tests to see what will react*. This is unnecessary and may be dangerous. Do *only* those tests indicted in Table 16-1 for the compound(s) you think may be present on the basis of color.

II. SPECIFIC EXPERIMENTS

Your instructor will assign selected experiments from the list below.

Experiment 1.   Place 4 mL of 0.1 M* $CuSO_4$ solution in a 15-cm test tube. Add 7 mL of 0.1 M KOH. Stir, and watch for any gas or precipitate. Record your observations.

Experiment 2.   Place a small marble chip ($CaCO_3$) in a 15-cm test tube. Add 6 mL of 6 M HCl. Observe and record. (*Note:* H_2O is one of the products.)

Experiment 3.   Place 6 mL of distilled H_2O in a 15-cm test tube. Add a small *lump* (not turnings) of Ca metal. Observe and record.

Experiment 4.   Place a small piece of Mn metal in a 15-cm test tube. Add 8 mL of 1.0 M $HClO_4$. Observe and record.

Experiment 5.   Place 3 mL of 0.2 M H_2O_2 in a test tube. Add 5 mL of 0.2 M $Ce(ClO_4)_4$. Observe and record. Shake test tube before testing for a gas.

Experiment 6.   Place 7 mL of 0.05 M $KCr(OH)_4$ in a 15-cm test tube. Add 5 mL of 0.1 M H_2SO_4. Stir and record. Make all measurements very carefully. (*Note:* H_2O is a product of the reaction.)

Experiment 7.   Place 3 mL of 0.03 M $MnSO_4$ in a 15-cm test tube. Add 3 mL of 0.02 M $KMnO_4$. Watch closely (and record) the color changes that occur, and then determine the final products. (*Note:* H_2O is one of the *reactants*—left-hand side of equation).

Experiment 8.   Place 6 mL of 0.2 M $Ce(ClO_4)_4$ in a test tube. Add 3 mL of 0.2 M $H_2C_2O_4$. Stir thoroughly. Observe and record.

Experiment 9.   Begin heating 200 mL of tap water in a 400-mL beaker (boiling stones!). In a 15-cm test tube, place 4 mL of 0.2 M H_2O_2 and 1 mL of 0.5 M KOH. Mix thoroughly. Now add 4 mL of 0.05 M $KCr(OH)_4$. Again, mix thoroughly. Place the test tube in the hot water bath, and heat until the tap water boils. Then you may shut off the heat and determine

*M = molarity, a unit of concentration.

the products of the reaction, *of which H₂O is one:* The bubbles of gas given off by the reaction during the heating are due to decomposition of the H_2O_2 and should be ignored.

Experiment 10. Place 4 mL of 0.1 *M* $CuSO_4$ and 3 mL of 0.1 *M* SO_2 in a 15-cm test tube. Stir thoroughly. Now add 4 mL of 0.1 *M* HI. Stir and record your observations. (*Note:* H_2O is a reactant.)

III. SPECIAL TESTS

Note: A fresh sample must be used for each test. If your sample contains liquid and solid, separate these before proceeding.

1. *Hydrogen, H₂.* To test for this gas, bring a *burning* splint up to the mouth of the test tube. A flash or audible "pop" shows the presence of H_2.
2. *Oxygen, O₂.* To test for this gas, plunge a *glowing* (not burning) splint into the test tube. If the splint glows brightly or ignites, O_2 is present.
3. *Carbon dioxide, CO₂, and nitrogen, N₂.* These gases do not burn and will extinguish a glowing splint.
4. *NO₂, Cl₂, and H₂S.* These gases are identified by their color and/or odor.
5. *Test with AgNO₃.* Place about 1 mL of the solution to be tested in a test tube, and add a few drops of 0.1 *M* $AgNO_3$. Watch for a precipitate.
6. *Test with BaCl₂.* Place about 1 mL of the solution to be tested in a test tube, and add a few drops of 0.1 *M* $BaCl_2$. Watch for a precipitate.
7. *Test with phenolphthalein (for bases).* To 1 mL of the solution to be tested, add a few drops of phenolphthalein solution. Watch for a pink color.
8. *Test for Congo Red Paper (for acids).* Place 1 drop of the solution to be tested on a strip of Congo Red paper. Watch for a dark blue color, which is a positive test.
9. *Test with CuSO₄ (for NH₃).* Place about 1 mL of the solution to be tested in a test tube, and add a few drops of 0.5 *M* $CuSO_4$. A purple color indicates that NH_3 is present.

EXPERIMENT 16

CHEMICAL REACTIONS
PRE-LABORATORY QUESTIONS

1. An old water pipe is clogged with deposits. A chip of the deposit is tested with HCl and found to cause evolution of a gas. A splint does not burn or glow in the gas. Venture an educated guess as to the identity of the deposit.

EXPERIMENT 16

CHEMICAL REACTIONS
REPORT SHEET

Name _____   Section _____

Experiment 1.   List all the products you think were formed. What evidence do you have for each

Write the balanced equation for the reaction that occurred.

Experiment 2.   List all the products you think were formed. What evidence do you have for each?

Write the balanced equation for the reaction that occurred.

Experiment 3.   List all the products you think were formed. What evidence do you have for each?

Write the balanced equation for the reaction that occurred.

Experiment 4.   List all the products you think were formed. What evidence do you have for each?

Write the balanced equation for the reaction that occurred.

Experiment 5. List all the products you think were formed. What evidence do you have for each?

Write the balanced equation for the reaction that occurred.

Experiment 6. List all the products you think were formed. What evidence do you have for each?

Write the balanced equation for the reaction that occurred.

Experiment 7. Describe the color changes that occurred.

List all the final products you think were formed. What evidence do you have for each?

Write the balanced equation for the reaction that occurred.

Experiment 8. List all the products you think were formed. What evidence do you have for each?

Write the balanced equation for the reaction that occurred.

Experiment 9. List all the products you think were formed. What evidence do you have for each?

Write the balanced equation for the reaction that occurred.

Experiment 10.   List all the products you think were formed. What evidence do you have for each?

Write the balanced equation for the reaction that occurred.

Problem

Hot solutions of NH_4Cl and KNO_2 are mixed. A colorless, odorless gas is evolved that does not burn and does not ignite a glowing splint. The clear, colorless solution that remains gives a precipitate with $AgNO_3$ but does not turn pink with phenolphthalein and does not turn Congo Red paper blue.

What products do you think were formed, and why? (H_2O is one of the products.)

Write the balanced equation for this reaction.

EXPERIMENT 17

SOLUTIONS: GENERAL STUDIES

Materials Needed

Salt, NaCl (10 g fine, 2 g coarse)
10 g sodium thiosulfate, $Na_2S_2O_3$
2 or 3 large crystals of cobalt(II) chloride, $CoCl_2$
Conductivity apparatus
3 mL 0.1 M $KMnO_4$

Background

Most chemical reactions in nature and in the laboratory take place not between compounds in the pure state but rather between compounds dissolved in water. The water simply allows the reagents to come into close contact with each other and thus react faster. It is therefore important that we study the properties of dissolved matter.

Solute is the term applied to the material to be dissolved, while the *solvent* is the substance that dissolves the solute. The resulting homogeneous mixture is the *solution*. In this experiment the common solutes will be salts and the common solvent will be water, but it should be remembered that solids, liquids, and gases can all be solutes or solvents. Thus, air is a solution of oxygen in nitrogen. Marine life depends on the solution of oxygen in water. Alloys are homogeneous solutions of two or more metals. Generally, in solutions the solute is the constituent present in the lesser amount.

A solution can obviously be very concentrated or dilute. The upper limit of concentration is a *saturated* solution that has taken up all the solute it can hold. Any additional solute will fail to dissolve and will remain separated from the solution unless it is heated. Except for gaseous solutes, heat will generally admit more solute into solution. The total amount of a particular solute that will dissolve in a particular solvent depends on the temperature of the solution.* Occasionally, when a hot, saturated solution is cooled carefully, the extra solute does not fall out. Such a solution is said to be *supersaturated*. Supersaturated solutions are

---

*For gases, solubility also depends on pressure.

generally on the verge of precipitating the extra solute. Precipitation of this extra solute may be caused by scratching the inside of the vessel with a glass rod or by introducing a crystal of the solute.

Solutes may dissolve into a solution quickly or slowly and in large or small quantities. Usually a solute will dissolve faster the smaller its particle size, the faster the stirring, the higher the temperature, and the more dilute the solution is.

But the *rate* of dissolving is independent of the *amount* of solute that will eventually dissolve. For example, coarse solutes will dissolve more slowly than powdered solutes. But given enough time (in other words, disregarding the rate), the total amount that will eventually dissolve is the same whether the solute is coarse or fine.

Procedure (You may be instructed to work in pairs)

I. SATURATED SOLUTION

Obtain a clean, dry 100-mL beaker. Weigh it and record its mass on the Report Sheet. Fill it one-fourth full with distilled water and weigh again. Now slowly add some fine NaCl crystals and stir the solution continuously with a stirring rod. Keep stirring and adding salt until no more dissolves and a few undissolved crystals are visible on the bottom of the dish. Again weigh and record the mass.

II. EFFECT OF TEMPERATURE

Place the saturated solution prepared above onto a ring stand. Stir and heat *gently*. Watch the crystals, and record your observations.

III. EVAPORATION

Continue heating the beaker to evaporate some solvent. The solution will get quite hot and may spatter occasionally. Heat, with stirring, until you are sure that the total volume has been reduced somewhat. Allow to cool and then observe.

IV. SUPERSATURATION

Set up a 400-mL beaker as shown in Fig. 1. Fill it about half full with water, and heat to a simmer.

Add 3 to 4 g of sodium thiosulfate to 3 mL of distilled water in a large test tube. Heat the test tube and its contents in the 400-mL beaker as shown in the figure. Stir the solution until all crystals are dissolved. Best mixing is accomplished by stirring with both a circular motion and an up-and-down motion. Once all crystals are dissolved, add a few more crystals and stir to dissolve. Repeat until no more solute dissolves. At this point add enough distilled water (5 drops at a time) with stirring to dissolve the remaining crystals. Put the test tube and its contents aside and allow to cool undisturbed for some 30 min. Go on with the next part of the experiment.

Add a small crystal of solid sodium thiosulfate to the cooled solution. Observe what happens, note any changes in appearance, temperature, etc., and record on the Report Sheet.

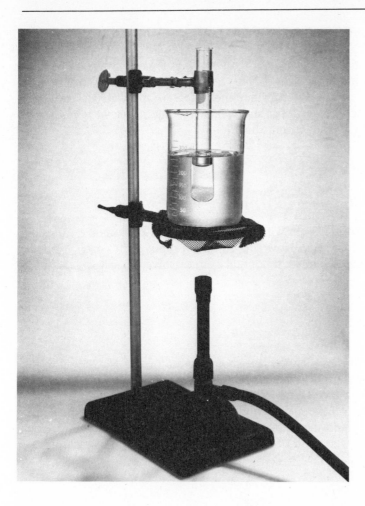

FIGURE 1 Assembly used to observe supersaturation. (*Courtesy of W. K. Fife.*)

V. SPEED OF DISSOLUTION VERSUS DIFFUSION

Place one large crystal of cobalt(II) chloride at the bottom of a test tube and slowly and carefully fill the test tube nearly full with distilled water. Set aside in a rack.

Nearly fill a second test tube with distilled water, and then place one large crystal of cobalt(II) chloride into a small funnel-folded filter paper and put it in the test tube as shown in Fig. 2. The end of the filter paper should be submerged in the water. Set aside in a rack.

Observe the two test tubes and filter paper occasionally over the next hour. Make sure that the filter paper always touches the solution in the test tube. Report the results.

FIGURE 2 Setup for cobalt(II) chloride diffusion.

VI. SPEED OF DISSOLUTION VERSUS TEMPERATURE

Weight out two 0.5-g samples of fine NaCl crystals. Now take two clean beakers (100-mL and 150-mL sizes are appropriate, but other sizes may be used) and add about 50 mL of cold tap water to one and about 50 mL of hot water to the other. Be sure the water is hot! Now add the 0.5-g samples of salt to each beaker, stir, and watch the time it takes to dissolve in each case. Record your observation on the Report Sheet.

VII. SPEED OF DISSOLUTION VERSUS PARTICLE SIZE

Fill a dry test tube to a depth of about 0.5 cm with fine NaCl crystals. Fill another dry tube to the same depth with coarse NaCl crystals. Now add 10 mL of tap water to each, stopper, and shake both tubes at the same time. Note the time it takes to dissolve the salt in each tube.

VIII. CONDUCTIVITY

Obtain two clean, dry evaporating dishes. To one add about a gram of fine NaCl, and to the other a few milliliters of distilled water. Test the conductivity of the solid dry salt by slowly allowing it to come into contact with the electrodes of a conductivity apparatus. See Fig. 3. (**Caution:** *Do not touch the electrodes! Be sure your dish and hands are dry!*) Now test the conductivity of the distilled water. With the electrodes dipping into the distilled water, pour the salt from the other dish into the water and observe the result. Record all observations.

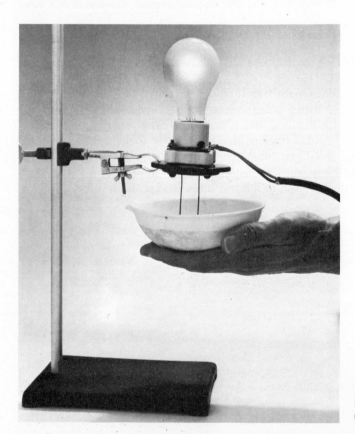

FIGURE 3 Conductivity apparatus. (*Courtesy of W. K. Fife.*)

IX. WASHING EFFICIENCY

This exercise is meant to convince you that *one washing using a large volume is not as efficient as several washings using the same total volume of solvent* [see also *J. Chem. Educ.*, *49*, 650 (1972)].

Obtain two clean 125-mL erlenmeyer flasks and label them A and B. Into each pour 1 mL of 0.1 *M* $KMnO_4$. The object now is to wash out the $KMnO_4$. Into flask A pour 90 mL of tap water, and into flask B pour 30 mL. Holding each flask in one hand, simultaneously swirl and pour out into the sink and allow to drip for a second or two. Then set them right side up. You will probably notice that B has more color left than A; however, it only has been rinsed with one-third as much solvent. Flask A may be set aside.

Now add a fresh portion of 30 mL of water to *flask B only*. Rinse and decant as before, allowing to drop for a second or two. Again, add 30 mL of fresh water and rinse as before. You have now used a total of 90 mL for each flask—but you have done so in different ways. Closely inspect the two flasks while holding them side by side. Look at the drops of liquid left. Which washing was more efficient? Keep this in mind!

EXPERIMENT 17

SOLUTIONS: GENERAL STUDIES: PRE-LABORATORY QUESTIONS

1. The solubility of cobalt(II) chloride in 100 mL of water is a linear relationship to temperature. Thus, 29.5 g dissolves at 0°C and 35.5 g at 30°C. What is the solubility of the salt at 15°C?

_____ g

2. The solubility of cane sugar in water at 0°C is 1.792 g/mL water. A 300-mL glass of iced tea is treated with 50 g of sugar. Is the resulting solution unsaturated or saturated? Show your work.

_____

3. Consider oxygen gas dissolved in oceans. Suggest the best depth (shallow or deep) and temperature (warm or cold) combination for maximum oxygen dissolution.

EXPERIMENT 17

SOLUTIONS: GENERAL STUDIES
REPORT SHEET

Name _____   Section _____

I. SATURATED SOLUTION

Weight of beaker + water ____.____ g   Weight of beaker + water + salt ____.____ g

Weighing beaker ____.____ g   Weight of beaker + water ____.____ g

Weight of water ____.____ g   Weight of salt ____.____ g

What is the concentration of your saturated salt solution in grams of salt per milliliter of solvent? [*Note:* Density of water = 1.00 g/mL.]

II. EFFECT OF TEMPERATURE

In part I, the solution was saturated with a small amount of solute left undissolved. As you heat the solution, what happens to the undissolved solute? In other words, what effect does an increase in temperature have on the total amount of solute in solution?

III. EVAPORATION

This is one method used in obtaining fresh water from seawater. The evaporation process drives off pure water and leaves the salt in the dish. What does this indicate about the boiling point of the salt compared to that of water?

IV. SUPERSATURATION

Describe and explain your observations when a crystal of sodium thiosulfate is added to the cooled solution.

V. SPEED OF DISSOLUTION VERSUS DIFFUSION

Describe and explain any differences in the two test tubes.

VI. SPEED OF DISSOLUTION VERSUS TEMPERATURE

Time Needed for Dissolution

Cold solution _____ minutes + _____ seconds

Hot solution _____ minutes + _____ seconds

Explain your observations.

VII. SPEED OF DISSOLUTION VERSUS PARTICLE SIZE

Time Needed for Dissolution

Fine salt _____ minutes + _____ seconds

Coarse salt _____ minutes + _____ seconds

Explain your observations.

VIII. CONDUCTIVITY

Substance	Does it Conduct Electric Current?
Dry salt	_____
Distilled water	_____
Salt in water	_____

Explain your observations.

IX. WASHING EFFICIENCY

Which washing method is more effective?

Problems

1. Tap water in most cities does conduct some electric current. Explain.

2. Honey is a supersaturated solution. What evidence proves this statement?

EXPERIMENT 18

METAL IONS IN ROCKS

Materials Needed

0.05 M AgNO$_3$
0.10 M Al(NO$_3$)$_3$
0.05 M Ca(NO$_3$)$_2$

0.05 M Cu(NO$_3$)$_2$
0.05 M Fe(NO$_3$)$_3$ in 0.1 M HNO$_3$
0.05 M Ti(IV) nitrate*

Small "dropper bottles" containing the following:
3 M KOH
3 M KCL
0.5 M K$_2$C$_2$O$_4$
3 M H$_2$SO$_4$
Phenolphthalein solution

3% fresh H$_2$O$_2$
3 M NH$_4$OH
3 M HCl
2 M H$_3$PO$_4$
Fresh 0.50 M NaI/0.15 M Na$_2$SO$_3$*

Minerals containing one or two of the following metals: Ag, Al, Ca, Cu, Fe, Ti
Unknowns corresponding to the metal ions in each of the minerals

Background

Although elemental metals are familiar to everyone and have many uses, they do not occur uncombined in nature, except in rare cases. Instead, metals are found combined with other elements in *minerals;* rocks are composed of varying mixtures of minerals, sometimes of many different metals. Table 18-1 lists the *average* abundance of some metals in the earth's crust, and in what minerals they are found.†

---

*Preparation of solutions:
0.50 M NaI/0.15 M Na$_2$SO$_3$: Dissolve 7.50 g of NaI and 1.80 g of anhydrous Na$_2$SO$_3$ in H$_2$O, and dilute to 100 mL. The solution does not keep well due to oxidation of Na$_2$SO$_3$.
0.05 M Ti(IV) nitrate in 1 M HNO$_3$: Fuse a mixture of 20 g of KOH and 5 mL H$_2$O (distilled) in a Ni crucible, by gentle heating (it may froth somewhat). Add 4.0 g of TiO$_2$, in portions. The mixture may be stirred occasionally and briefly with a stainless steel spatula—do not use glass or aluminum. After the TiO$_2$ has been added, heat the crucible (Fisher burner) at red heat for at least 15 min. It is not necessary to cover the crucible, and the light brown contents should be stirred or poked occasionally to prevent sticking and caking. Allow to cool, but grind up the contents in a mortar while still warm—otherwise it becomes very hard. Add the resulting powder in portions to a mixture of 100 mL of distilled H$_2$O and 300 mL of 6 M HNO$_3$ (*Caution: Heat evolved. Stir thoroughly, filter if necessary, and dilute to 1 liter with distilled H$_2$O*).

†Seawater also contains metals, as their dissolved ions, and is a commercial source of sodium and magnesium.

Table 18-1   Abundance of some metals in the earth's crust

Metal	Abundance per lb of Rock	Representative Minerals
Al	37 g	Al_2O_3
Fe	23 g	Fe_2O_3
Ca	17 g	$CaCO_3$, $CaSO_4$
Na	13 g	$NaCl$, $NaNO_3$
Mg	9.5 g	$MgCO_3$
Ba	1.1 g	$BaCO_3$, $BaSO_4$
Cu	30 mg	Cu_2O, $CuCO_3$
Pb	7.0 mg	PbO, PbS
Dy	2.0 mg	$DyPO_4$
Hg	0.20 mg	HgS
Bi	0.09 mg	Bi_2S_3, "free" Bi
Ag	0.05 mg	Ag_2S, "free" Ag
Au	0.002 mg	"Free" Au

It should be noted that some "obscure" metals such as Dy (dysprosium) are more abundant than some of the more familiar ones.

Though all of these minerals contain metal ions, rocks obviously are not very soluble in water; therefore, the ionic compounds found in these minerals must not be soluble, since soluble minerals would get washed into the ocean. Soluble minerals do indeed exist, such as saltpeter (KNO_3), but only in very dry desert areas or in underground deposits.

Since you have previously learned "ionic compounds are usually soluble in water," how do you know which ones are exceptions? Note in Table 18-1 that oxides and sulfides often are representative minerals, suggesting that oxides and sulfides of many metals are insoluble, which is the case. Table 18-2 gives a more complete listing of solubility rules.

What is the chemical significance of solubility? If the components of an insoluble compound "meet" in aqueous solution, a solid *precipitate* will form. To illustrate this, consider the following reaction:

$$BaCl_2(aq) + Na_2SO_4(aq) \rightarrow ?$$

First, both $BaCl_2$ and Na_2SO_4 are soluble, since they are given as already in solution: "(aq)" = aqueous; see also solubility rules below.

Table 18-2   Solubility rules for ionic compounds in water

All metallic oxides, sulfides, and hydroxides are *insoluble* in water except those of groups IA and IIA.

All metallic carbonates and phosphates are *insoluble* except those of group IA.

All metallic nitrates and acetates are *soluble*.

All group IA and NH_4^+ (ammonium) salts are *soluble*.

All metallic sulfates are soluble *except* those of Ca^{2+}, Sr^{2+}, Ba^{2+}, Pb^{2+}, and Hg^+; Ag_2SO_4 is slightly soluble.

All metallic chlorides, bromides, and iodides are soluble *except* those of Pb^{2+}, Hg^+, Ag^+, and Cu^+; in addition, BiI_3 and HgI_2 are insoluble.

BaCl$_2$ (aq) contains Ba^{2+} (aq) and 2 Cl$^-$ (aq); Na$_2$SO$_4$ (aq) contains 2 Na$^+$ (aq) and SO$_4^{2-}$ (aq). These four ions are free to mix in solution. When Ba^{2+} (aq) "meets" SO$_4^{2-}$ (aq), a precipitate of BaSO$_4$ will form:

$$Ba^{2+}(aq) + SO_4^{2-}(aq) \rightarrow BaSO_4 \downarrow$$

Note in Table 18-2 that metallic sulfates are soluble *except* for those of Ca^{2+}, Sr^{2+}, Ba^{2+}, Pb^{2+}, and Hg$^+$.

Therefore, if aqueous BaCl$_2$ and Na$_2$SO$_4$ are mixed:

$$BaCl_2(aq) + Na_2SO_4(aq) \rightarrow BaSO_4 \downarrow + 2\,NaCl(aq)$$

You do not "see" that NaCl, which remains in solution. As far as BaSO$_4$ in nature is concerned, you might well find it in a mineral, but not in the ocean.

The "exchange" of ions is called *metathesis* or *double displacement*. All metathesis reactions take the form AX + BY → AY + BX where A and B are cations and X and Y are anions. You should *not* propose a formula such as BaNa$_2$ in the above BaCl$_2$ + Na$_2$SO$_4$ reaction because Na$^+$ and Ba^{2+} are both positively charged and will not form a compound.

Now consider:

$$CuSO_4(aq) + KOH(aq) \rightarrow$$

To see what, if any, products might be formed, exchange the ions; this gives Cu(OH)$_2$ and K$_2$SO$_4$. Beware! Do *not* write "CuOH" or "KSO$_4$." Chemical formulas must *not* be changed so that a chemical equation will balance.

Since all Group IA compounds are soluble (see Table 18-2), K$_2$SO$_4$ is soluble. However, Cu(OH)$_2$, a metallic hydroxide, is insoluble and will precipitate:

$$CuSO_4(aq) + 2\,KOH(aq) \rightarrow Cu(OH)_2 \downarrow + K_2SO_4(aq)$$

It is essential to understand that *predicting* the products of a reaction is not the same process as balancing an equation.

Another example:

$$NaCl(aq) + MgSO_4(aq) \rightarrow$$

Exchange or metathesis of these ions would give Na$_2$SO$_4$ and MgCl$_2$. From Table 18-2 you should be able to deduce that both of these are soluble—no precipitate can be formed. Therefore, if aqueous NaCl (table salt) and aqueous MgSO$_4$ (Epsom salts) were mixed, nothing would happen (NR = no reaction).

Metathesis reactions form much of the basis of a technique known as *qualitative analysis*. This is the application of chemical principles in the laboratory in identification of the elemental composition of substances. Traditionally, these "substances" have included anions and cations in aqueous solution, as well as simple salts such as NiCl$_2$ and ZnSO$_4$.

One of the earliest chemists to organize analytical reactions was Robert Boyle (1627–1691), who was the first to use litmus (a vegetable dye) to detect acids and bases. Later, metal cations were separated into groups by the Swedish chemist T.O. Bergman (1735–1784). In the following centuries, coherent systematic methods—"schemes"—of analysis gradually emerged. In 1871, N. A. Menshutkin published *Analytical Chemistry,* and in the next 50 years many other such works followed. Until the middle of this century, these analytical schemes were among the primary methods of identification of metals, nonmetals, and their associated ions. Today, instrumental methods are faster and can identify trace amounts not detectable by chemical methods. However, the use of such instruments appears "exotic" and does not illustrate chemical properties.

For the purpose of this experiment, you will work with six cations: Ag$^+$, Al^{3+}, Ca^{2+},

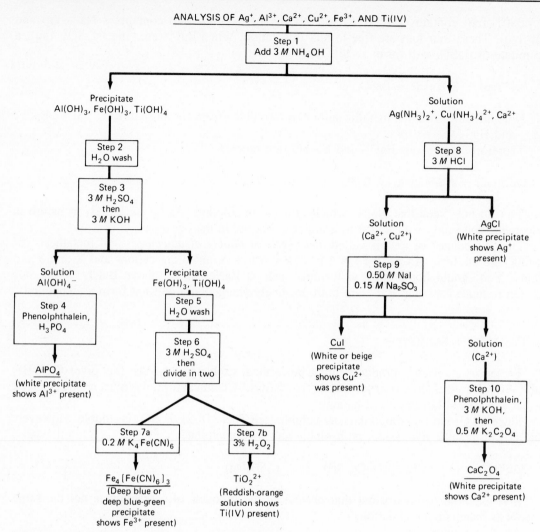

ANALYSIS OF Ag^+, Al^{3+}, Ca^{2+}, Cu^{2+}, Fe^{3+}, AND Ti(IV)

Step 1
Add 3 M NH_4OH

Precipitate
$Al(OH)_3$, $Fe(OH)_3$, $Ti(OH)_4$

Step 2
H_2O wash

Step 3
3 M H_2SO_4
then
3 M KOH

Solution
$Al(OH)_4^-$

Step 4
Phenolphthalein,
H_3PO_4

$AlPO_4$
(white precipitate
shows Al^{3+} present)

Precipitate
$Fe(OH)_3$, $Ti(OH)_4$

Step 5
H_2O wash

Step 6
3 M H_2SO_4
then
divide in two

Step 7a
0.2 M $K_4Fe(CN)_6$

$Fe_4[Fe(CN)_6]_3$
(Deep blue or
deep blue-green
precipitate
shows Fe^{3+} present)

Step 7b
3% H_2O_2

TiO_2^{2+}
(Reddish-orange
solution shows
Ti(IV) present)

Solution
$Ag(NH_3)_2^+$, $Cu(NH_3)_4^{2+}$, Ca^{2+}

Step 8
3 M HCl

Solution
(Ca^{2+}, Cu^{2+})

AgCl
(White precipitate
shows Ag^+
present)

Step 9
0.50 M NaI
0.15 M Na_2SO_3

CuI
(White or beige
precipitate
shows Cu^{2+}
was present)

Solution
(Ca^{2+})

Step 10
Phenolphthalein,
3 M KOH,
then
0.5 M $K_2C_2O_4$

CaC_2O_4
(White precipitate
shows Ca^{2+} present)

FIGURE 1   Separation and identification of some metal ions.

Cu^{2+}, Fe^{3+}, and Ti^{4+}. These of course are not the only possible ions that can be identified, nor is the scheme used in this experiment the only known. The ions selected not only represent a wide range of properties in the periodic table, but also have commercial, geological, or biological significance. Al^{3+} and Fe^{3+} are the first and second most abundant metal ions in the earth's crust. In addition, Al and Fe as elements have major structural uses, as does Ti in certain alloys. Ag is valued in jewelry, photography, and (formerly) in coins. Ca^{2+} and Cu^{2+} are biologically essential in the human body.

Before analyzing an unknown, you will first do preliminary tests with the six ions in aqueous solution so that you will understand metathesis reactions. Though oxalate ion ($C_2O_4^{2-}$) is not listed in the solubility rules, you will deduce its properties through reaction with each of the metal ions.

You will then choose a mineral, and will be given an unknown containing ions (up to three) present in the mineral.

The cation analysis scheme outline in Fig. 1 should be consulted frequently so that you have an overview of the experiment.

Procedure

Bear in mind the formation of slight cloudiness may be due to impurities, dirty test tubes, or poor technique. *Stir* solutions after adding any chemical reagents. After measuring a volume of any metal ion in a graduated cylinder, rinse the graduated cylinder with distilled water before going on to the next metal ion.

I. METATHESIS REACTIONS

1. Silver ion, Ag^+. Place 4 mL 0.05 M $AgNO_3$ in each of three test tubes.
    To the first, add 4 drops of 3 M KOH.
    To the second, add 4 drops of 3 M KCl.
    To the third, add 10 drops of 0.5 M $K_2C_2O_4$. (*Caution! Toxic!*)

Record the appearance of each precipitate (if any) and write a balanced chemical equation *if* a precipitate was formed. Do not discard the silver residues. A container will be provided.

2. Aluminum ion, Al^{3+}. Place 4 mL 0.10 M $Al(NO_3)_3$ in each of three test tubes.
    To the first, add 3 drops of 3 M KOH.
    To the second, add 6 drops 3 M KCl.
    To the third, add 10 drops 0.5 M $K_2C_2O_4$.

Record your observations. In each case where a precipitate was formed, write a balanced chemical equation. Aluminum residues may be washed down the sink.

3. Calcium ion, Ca^{2+}. Place 4 mL 0.05 M $Ca(NO_3)_2$ in each of three test tubes.
    To the first, add 6 drops 3 M KCl.
    To the second, add 3 drops 3 M H_2SO_4.
    To the third, add 10 drops 0.5 M $K_2C_2O_4$.

Record your observations. In each case where a precipitate was formed, write a balanced chemical equation. Calcium residues may be washed down the sink.

4. Copper ion, Cu^{2+}. Place 4 mL 0.05 M $Cu(NO_3)_2$ in each of three test tubes.
    To the first, add 4 drops 3 M KOH.
    To the second, add 6 drops 3 M KCl.
    To the third, add 10 drops 0.5 M $K_2C_2O_4$.

Record your observations. In each case where a precipitate was formed, write a balanced chemical equation. Copper residues may be washed down the sink.

5. Iron ion, Fe^{3+}. Place 4 mL of 0.05 M $Fe(NO_3)_3$ in each of three test tubes.
    To the first, add 8 drops 3 M KOH.
    To the second, add 8 drops 3 M KCl.
    To the third, add 12 drops 0.5 M $K_2C_2O_4$.

Record your observations. In each case where a precipitate was formed, write a balanced chemical equation. Iron residues may be washed down the sink.

6. Titanium ion, Ti^{4+}. Place 4 mL 0.05 M Ti(IV) nitrate solution in each of three test tubes.
    To the first, add 25 drops 3 M KOH
    To the second, add 25 drops 3 M KCl.
    To the third, add 25 drops 0.5 M $K_2C_2O_4$.

Record your observations. In each case where a precipitate was formed, write a balanced chemical equation. Titanium residues may be washed down the sink.

II. PHYSICAL STUDY OF METAL ION UNKNOWN

Choose a rock or mineral from the available selection. Describe as many of its physical characteristics as possible (color, shininess). Do not break or chemically test the sample. Write down the unknown number of your sample. You will be given a solution containing the metal ion(s) present in the sample.

III. CHEMICAL STUDY OF METAL ION UNKNOWN

Your rock or mineral contains one to three of the six ions in this experiment. It may be helpful to refer to the flowsheet (Fig. 1) and to label your test tubes according to the ions that may be present in each step.

Known solutions containing the individual six ions, as well as all six together, are available. When you think you are sure of the identity of your unknown, obtain a known solution containing these ions, and confirm your answers by use of the known solution. In addition, the known solutions may be used to confirm the color of a precipitate.

Caution: Use distilled water throughout!

Step 1.   Place 40 drops of the unknown in a clean 15-cm test tube. Add 45 drops of 3 M NH_4OH, and stir well. Centrifuge and precipitate (that's the solid) that is formed, decant, and save the supernatant (that's the liquid) for step 8. If no precipitate forms, this means the ions Al^{3+}, Fe^{3+}, and Ti(IV) are all absent—skip steps 2 to 7 and proceed to step 8. Otherwise, proceed with the precipitate to step 2.

Step 2.   Wash the precipitate from step 1 by adding 5 mL distilled H_2O, mixing well, and centrifuging. Discard the supernatant, and save the precipitate for step 3.

Step 3.   Add 40 drops of 3 M H_2SO_4 to the precipitate from step 2, and stir or shake until it all dissolves. Then add 6 mL (not drops) of 3 M KOH. Stir well, centrifuge, and decant. Save the precipitate for step 5 and the supernatant for step 4. If no precipitate was formed, Fe^{3+} and Ti(IV) are absent; proceed to step 4, but skip steps 5 to 7.

Step 4.   Add 1 drop of phenolphthalein solution (an indicator that shows acidity or basicity of your solution). If this does not turn the solution pink, add drops of 3 M KOH (with stirring) until it is pink (indicates basic). Now add 2 M H_3PO_4 dropwise (with stirring) until the pink color disappears (shows slightly acidic solution). Centrifuge. A white precipitate is $AlPO_4$ and shows Al^{3+} is present. (If no precipitate forms, Al^{3+} is absent.) Proceed to step 5.

Step 5.   Wash the precipitate (from step 3, not 4) by adding 5 mL of distilled H_2O, stirring, and centrifuging. Discard the supernatant, and save the precipitate. Proceed to step 6.

Step 6.   Add 40 drops of 3 M H_2SO_4 to the precipitate from step 5. Stir until dissolved; then divide the solution into two roughly equal parts. Proceed to steps 7a and 7b.

Step 7a.   To one of the portions, add 10 drops of 0.2 M $K_4Fe(CN)_6$. Centrifuge. A deep blue or deep blue-green precipitate is $Fe_4[Fe(CN)_6]_3$ and shows that Fe^{3+} is present.

Step 7b.   To the other portion, add 10-20 drops of 3% H_2O_2. Do not centrifuge. A reddish-orange solution is due to the TiO_2^{2+} ion and shows that Ti(IV) is present.

Step 8.   Add 40 drops of 3 M HCl to the solution you saved from step 1. Centrifuge, decant, and save the supernatant for step 9. A white precipitate is AgCl and shows that Ag^+ is present. (If no precipitate forms, no Ag^+ is present.)

Step 9.   Add 30 drops of 0.50 M NaI/0.15 M Na_2SO_3 solution. Centrifuge and save the supernatant for step 10. A white or beige precipitate is CuI and shows that Cu^{2+} was present. (If no precipitate forms, no Cu^{2+} is present.)

Step 10.   Add 1 drop of phenolphthalein solution, then add 3 M KOH dropwise with stirring until a light pink color remains (due to phenolphthalein—shows slightly basic solution). Now add 20 drops of 0.5 M $K_2C_2O_4$ (**Caution:** *toxic*). A white precipitate is CaC_2O_4 and shows that Ca^{2+} is present.

EXPERIMENT 18

METAL IONS IN ROCKS
PRE-LABORATORY QUESTIONS

1. Of these six ions, which ion is present when you have
    (a) A blue solution?

    (b) A colorless solution which forms a white precipitate when a solution of Cl^- ion is added?

2. In step 4 of the flowchart what ion is identified upon the addition of phenolphthalein and H_3PO_4?

EXPERIMENT 18

METAL IONS IN ROCKS
REPORT SHEET

Name _____   Section _____

I. METATHESIS REACTIONS

1. Silver ion, Ag^+. For each reaction undergone by Ag^+ which gave a precipitate, describe the precipitate and write a balanced chemical equation.

2. Aluminum ion, Al^{3+}. For each reaction undergone by Al^{3+} which gave a precipitate, describe the precipitate and write a balanced chemical equation.

3. Calcium ion, Ca^{2+}. For each reaction undergone by Ca^{2+} which gave a precipitate, describe the precipitate and write a balanced chemical equation.

4. Copper ion, Cu^{2+}. For each reaction undergone by Cu^{2+} which gave a precipitate, describe the precipitate and write a balanced chemical equation.

5. Iron ion, Fe^{3+}. For each reaction undergone by Fe^{3+} which gave a precipitate, describe the precipitate and write a balanced chemical equation.

6. Titanium ion. Though solutions of titanium(IV) ion are quite complex and contain ions such as $TiCl_6^{2-}$ = and $Ti(OH)_2^{2+}$, you may assume that Ti^{4+} is present as $Ti(NO_3)_4$ for the purpose of writing equations for each precipitate obtained.

II. PHYSICAL STUDY OF METAL ION UNKNOWN (MINERAL)

1. Record the unknown number _____.

2. Describe the physical properties of the rock or mineral you chose.

III. CHEMICAL STUDY OF METAL ION UNKNOWN

What ions(s) did you find? Give *evidence*. It is not necessary to write chemical equations.

Problems

1. When an unknown solution containing one or two metal ions was treated with a solution containing HCl and H_2O_2, a precipitate and a reddish-orange solution was formed.

    (a) What color is the precipitate? What is it?

    (b) What is the reddish-orange solution due to?

2. Distinguish between the following pairs of ions, using one chemical reagent. Describe what you would *see* and *do*.

    (a) Ca^{2+} and Fe^{3+}.

    (b) Cu^{2+} and Ag^+.

EXPERIMENT 19

ACIDS, BASES, AND SALTS

Materials Needed

Universal indicator
1 g each of P_2O_5, Al_2O_3, CaO, and SiO_2
4 g BaO
5.0 g KBr
10 mL concentrated H_3PO_4
A little white vinegar, household ammonia,
liquid bowl cleaner, baking powder, sal-soda,
Drāno, other commercial products
Red cabbage leaves or flower petals
20 mL 95% ethyl alcohol
0.1 M HCl and 0.1 M NaOH in
dropping bottles

Phenolphthalein indicator
Standardized NaOH (about 0.5 M)
Standardized HCl (about 0.5 M)
Dilute NaOH
5.0 g $NiCO_3$
20 mL 3 M H_2SO_4
Jar of NH_3 gas, covered with glass plate
Jar of HCl gas, covered with glass plate
Universal indicator

Background

The sour tastes of vinegar and lemons are due to acids—acetic and citric, respectively. Properties of acids and bases were known long before these properties could be explained or defined. A useful definition of an *acid,* proposed in 1923 by Brønsted and Lowry, is a substance which can donate H^+ (hydrogen ion), and a *base* is a substance which can accept H^+. Pure, neutral H_2O contains 10^{-7} M H^+ and 10^{-7} M OH^- from the dissociation*

$$H_2O \rightleftarrows H^+ + OH^-$$

An acidic substance will increase this hydrogen ion concentration to higher values. Common acids include HCl, HNO_3, H_2SO_4, H_3PO_4, and certain organic compounds called *carboxylic acids,* of which acetic and citric acids are examples. All organic acids contain the carboxylic acid group—COOH.

*In this discussion, H^+ will be used instead of H_3O^+ for the sake of simplicity.

The acidity of HNO_3 (nitric acid) can be illustrated in its reaction with H_2O:

$$HNO_3 + H_2O \rightarrow H_3O^+ + NO_3^-$$

The HNO_3 molecule, a strong acid, *dissociates* nearly completely in H_2O to give H_3O^+ (hydronium ion) and NO_3^- (nitrate ion). Note that HNO_3 fulfills the definition of an acid by donating H^+ to the H_2O. Similar reactions can be written for other acids.

Common bases include NaOH, KOH, $Mg(OH)_2$, $Ca(OH)_2$, NH_3, and amines (organic derivatives of NH_3). In H_2O, KOH, a white, ionic solid dissociates nearly completely:

$$KOH \rightarrow K^+(aq) + OH^-(aq)$$

Thus, KOH is a strong base. $Ca(OH)_2$, though of limited solubility, also dissociates in H_2O:

$$Ca(OH)_2 \rightarrow Ca^{2+}(aq) + 2\,OH^-(aq)$$

OH^- (hydroxide ion) fulfills the definition of a base since it reacts with any source of H^+:

$$H^+(aq) + OH^-(aq) \rightarrow H_2O$$

The reaction of an acid with a base in aqueous solution is called *neutralization*.

Though water is a common solvent for acid-base reactions, the presence of water is not required. Dry, gaseous HCl, and NH_3 will react to form the ionic solid NH_4 CL (ammonium chloride.) "Household ammonia" is a *solution* of NH_3 gas in water.

Most acidic and basic systems encountered in nature are low in concentration (see Table 19-1). To avoid the problem of having to use very small numbers to express such values, it has been found convenient to use the pH scale. It is defined as

$$pH = -\log[H^+]$$

which says that as the concentration (molarity) of H^+ increases, pH decreases, and consequently as the $[OH^-]$ increases, so does pH. The pH of 10^{-3} M H^+ is 3, the pH of 10^{-10} $[H^+]$ is 10, and that of 10^{-3} $[OH^-]$ is 11, since pH + pOH = 14.

There are several methods by which to measure the pH of a solution. The pH meter is an electronic instrument that gives direct and accurate pH readings. Certain organic compounds can also be used to indicate approximate acid (or base) strengths. These *indicators* change structure when the pH is changed, causing a change in color. A common commercially available indicator is *phenolphthalein,* whose structure is

(colorless)   (red)

Phenolphthalein

Universal indicators are a mixture of several commercial indicators. Naturally occurring indicators are present in some plants (for example, in red cabbage leaves, flowers, etc.) and may be extracted by boiling with H_2O or ethyl alcohol. Seasonal changes in the color of leaves are also due to changes in pH. The compound anthocyanin, found in the soft-maple leaf as well as in cranberries, is red in acid and purple in base.

Table 19-1 Some familiar acids and bases

Substance	Acid or Base Present	Approx. pH	[H$^+$]	[OH$^-$]
Stomach acid	HCl	1	10^{-1} M	10^{-13} M
Lime juice	Citric acid	2	10^{-2} M	10^{-12} M
Vinegar	Acetic acid	3	10^{-3} M	10^{-11} M
Milk of magnesia	Mg(OH)$_2$	9	10^{-9} M	10^{-5} M
Household ammonia	NH$_3$	11	10^{-11} M	10^{-3} M

Acids and bases are synthesized in a variety of ways. The binary acids HX (where X = F, Cl, Br, or I) result from heating the respective elements together but are more often prepared by displacement reactions, using the Na$^+$, K$^+$, or Ca^{2+} salt of the desired acid and concentrated H$_3$PO$_4$ or H$_2$SO$_4$. For example,

$$NaI + H_3PO_4 \rightarrow HI + NaH_2PO_4$$

or

$$\underset{\text{Sodium acetate}}{CH_3COO^-NA^+} + H_2SO_4 \rightarrow \underset{\text{Acetic acid}}{CH_3COOH} + NaHSO_4$$

HF (hydrofluoric acid), which is used to etch glass, is prepared by displacement directly from the mineral *fluorspar:*

$$CaF_2 + H_2SO_4 \rightarrow 2HF + CaSO_4$$

Nonmetallic oxides usually give acids in contact with H$_2$O:

$$SO_3 + H_2O \quad \rightarrow \underset{\text{Sulfuric acid}}{H_2SO_4}$$

$$B_2O_3 + 3H_2O \rightarrow \underset{\text{Boric acid}}{2H_3BO_3}$$

Conversely, *metallic* oxides of Group IA and IIA elements yield hydroxide ions in contact with H$_2$O, and are therefore basic:

$$Li_2O + H_2O \rightarrow \underset{\text{Lithium hydroxide}}{2LiOH}$$

Oxides of other metals are too insoluble in H$_2$O to significantly alter the pH. Aqueous ammonia is also a base, and ammonia may be prepared by displacement from ammonium salts by heating with a stronger base, such as NaOH or Ca(OH)$_2$:

$$NaOH + NH_4Cl \rightarrow NaCl + NH_3 + H_2O$$

Neutralization is the reaction of an acid with a stoichiometric amount of base, yielding a salt. The point at which the acid and base are neutralized is signaled by indicators and is referred to as the *endpoint*. These reactions are often very fast and complete and can be used for quantitative purposes. Such *titrations* depend on accurate measurement of the reactants involved. For instance, the reaction

$$H_2SO_4 + 2NaOH \rightarrow \underset{\text{(a salt)}}{Na_2SO_4} + 2H_2O$$

becomes a neutralization titration if exactly twice as many moles of NaOH are used as H_2SO_4. The following reaction, on the other hand, calls for a 1:1 ratio of reactants in order to be neutralized:

$$LiOH + CH_3COOH \rightarrow CH_3COO^-Li^+ + H_2O$$
$$\text{(a salt)}$$

The number of moles of acid (or base) is related to the mass of the pure substance, or the volume of its solution:

For pure substances:

$$\text{Moles} = \frac{\text{mass of substance}}{\text{molecular weight of substance}}$$

For substances in solution:

$$\text{Moles} = \text{molarity} \times \text{liters of solution}$$

Reaction of acid and base in aqueous solution can be used to prepare *salts:*

$$Ca(OH)_2(aq) + 2HNO_3(aq) \rightarrow Ca(NO_3)_2(aq) + 2H_2O$$

$$NaOH(aq) + H_3PO_4(aq) \rightarrow Na_3PO_4(aq) + 3H_2O$$

$NaPO_4$ is used as water softener and can be classified as a salt derived from the base NaOH and the acid H_3PO_4.

The reaction of metallic hydroxides with aqueous acids to give salts is a general one. Even though most metallic hydroxides are too insoluble in water to give very many OH^- ions, they behave as bases in reaction with acids, to give salts. These salts are generally ionic and soluble in water, so they are obtained as aqueous solutions. In order to obtain the solid salt, the water must be evaporated or, in some cases, the solution cooled.

$$Al(OH)_3 + 3HCl(aq) \rightarrow AlCl_3(aq) + 3H_2O$$

$$2HCl(aq) + Ni(OH)_2 \rightarrow NiCl_2(aq) + 2H_2O$$

Aluminum chloride is used in deodorants, nickel chloride in electroplating.

A reaction closely related to acid-base is that of metallic oxides and carbonates with acids to give a salt, water, and CO_2 (with carbonates):

$$Fe_2O_3 + 6HNO_3(aq) \rightarrow 2Fe(NO_3)_3(aq) + 3H_2O$$

$$ZnCO_3 + H_2SO_4(aq) \rightarrow ZnSO_4(aq) + H_2O + CO_2(gas)$$

In this experiment, you are to explore the indicator properties of some natural substances and test the acidity of some commercial products and of substances you prepare. You will then carefully titrate one or two of the aqueous solutions you prepared in order to calculate the molarity.

By way of illustrating such a calculation, assume that a 30.0-mL sample of HBr solution was titrated with standardized NaOH. The initial buret reading for the NaOH solution was 0.60 mL, and the final reading was 27.3 mL. The volume of base used was therefore 26.7 mL, or 0.0267 liter. The concentration of the standardized base is shown on the bottle to be 0.512 M. Thus, the number of moles used was:

$$0.0267 \text{ liter} \times 0.512 \text{ mole/liter} = 0.0137 \text{ mole NaOH}$$

This is a neutralization reaction (meaning all acid is neutralized with just enough base), and each acid molecule, HBr, contains one H^+, just as each base molecule, NaOH, contains one OH^-. Therefore, the 0.0137 mole of NaOH *must be equal* to the number of moles of acid initially present in the beaker: 0.0137 mole HBr. Then, the molarity of HBr may be calculated this way:

$$M = \frac{\text{moles of HBr}}{\text{liter of HBr solution}} = \frac{0.0137}{0.0300} = 0.457\ M$$

Procedure (May be done in groups, or spread over two laboratory periods)

I. SYNTHESIS

Note carefully: This experiment uses two different indicators and two different solutions of NaOH. Do not confuse them!

1. *Acids and bases from oxides.* In each of four 15-cm test tubes, place 5 mL of distilled H_2O. To the first tube, add a small amount of Al_2O_3, approximately that which will fit on the end 0.5 cm of a scoopula. Using the other test tubes, do the same for P_2O_5 (*Caution: Vigorous reaction!*) CaO, and SiO_2. Also prepare a fifth test tube with just distilled water. Stir the test tube with Al_2O_3 thoroughly. For the other test tubes, shake gently to mix thoroughly. Add a drop or two of universal indicator to each, and estimate the pH, using Table 19.2.

   Record your results on the Report Sheet. Account for the acidity or basicity of each substance.
2. *Barium hydroxide solution, Ba(OH)₂.* Add 4.0 g of BaO (*Caution: Toxic!*) to 100 mL of distilled H_2O in a 250-mL beaker, and stir thoroughly (several minutes). The solid will not all dissolve; gravity-filter the mixture, collecting the filtrate in a 125- or 250-mL erlenmeyer flask. *Save the filtrate for part III* (it may become slightly cloudy on standing).
3. *Hydrobromic acid, HBr.* Weigh 5.0 g of KBr, and place it in a dry 250-mL erlenmeyer flask. Add a few boiling stones and 10 mL of concentrated H_3PO_4 (*Caution: Corrosive!*) and swirl to mix. Assemble the apparatus shown in Fig. 1 (*hood*) and have it checked before proceeding. The entry tube in the second flask must *not* dip into the H_2O.

   Heat gently with a bunsen burner; the KBr + H_3PO_4 mixture will begin to bubble, and fumes will appear in the flask containing the H_2O. Continue heating until the KBr + H_3PO_4 mixture has become thick and syrupy, which should not be more than 10 minutes. Remove

Table 19-2   Approximate colors of universal indicator

pH	Color
2	Red
4	Orange
6	Yellow
8	Green
10	Blue

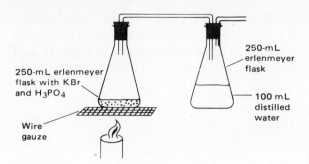

250-mL erlenmeyer
flask with KBr
and H_3PO_4

Wire
gauze

250-mL
erlenmeyer
flask

100 mL
distilled
water

FIGURE 1 Preparation of HBr.

the heat and disassemble the apparatus. Swirl the flask containing the H_2O, which is now an aqueous solution of HBr. Save this for part IV. The residue in the reaction flask should be rinsed *thoroughly* with H_2O. If this is sufficient to remove all gummy material, it may be soaked in NaOH solution (3 to 6 M).

4. *Synthesis of a salt.* Weigh 5.0 g $NiCO_3$, using a weighing boat, and place the $NiCO_3$ in a 125-mL flask. Add 20 mL 3 M H_2SO_4; some fizzing will begin (what is the gas?). Warm the mixture on a hot plate until all the solid has dissolved. Do not boil! Remove the flask from the hot plate, and add (*Caution: Flammable!*) 15 mL 95% ethyl alcohol. The purpose of the alcohol is to aid formation of crystals of the salt, since salts are usually less soluble in alcohols than in water. When the flask has cooled to near room temperature, place it in an ice bath for at least 15 min (while you are waiting, go on to other parts of the experiment). Suction-filter the crystals, and weigh them, again using a weighing boat. Record the appearance of the crystals; they will be collected, as will any waste solution in the filter flask.

II. pH OF COMMERCIAL PRODUCTS

Measure the pH of tap water, distilled water, white vinegar, household ammonia, and liquid bowl cleaner by adding one or two drops of universal indicator, *not* phenolphthalein, to 2 to 3 mL of each liquid in a clean test tube. Measure the pH of each of the following solids by dissolving about 0.1 g of each in 10 mL of distilled H_2O: baking powder, Sal-Soda, Drāno, and any commercial household products available such as milk of magnesia, pickle juice, dry wine, an orange, tomato, grape, etc.

III. GAS-PHASE ACID-BASE REACTION

Jars of NH_3 gas and HCl gas will be available. In the hood, carefully invert the jar of HCl over the NH_3, and remove the glass plates. What do you observe? What substance is formed? Is it soluble in water? Find out.

IV. ACID-BASE REACTIONS: TITRATIONS

1. *Naturally occurring indicators.* A variety of plant materials can be used for this section, including red cabbage leaves and numerous flower petals. Push the material to be tested to the bottom of a 15-cm test tube. If cabbage is used, add 3 mL of distilled H_2O; otherwise, use 3 mL of 95% ethyl alcohol. (*Caution: Flammable!*) Boil *gently* over a flame for 1 min, when a colored extract may be obtained. Add an equal volume of distilled H_2O, shake to mix, and divide it into two portions. To one, add drops of 0.1 M HCl, and to the other add drops of 0.1 M NaOH. Observe and record any color in each case.

2. *Titration of hydrobromic acid (HBr)*. Using a graduated cylinder or pipet remove 30.0-mL samples of your HBr from part I into two clean erlenmeyer flasks labeled 1 and 2. Clean our a buret by washing with detergent and rinsing several times with tap water and finally with distilled water. Each time allow some of the washings to drain through the tip, but then, with the stopcock tip open, turn the buret upside down into the sink to speed up the drainage.

　Obtain (using a clean, dry beaker) about 100 mL of the *standardized NaOH solution* (not 0.1 *M* NaOH) from the storeroom. Add 5 mL of it to the buret, and allow to drain. Repeat.

　Now fill the buret to above the 0.00-mL mark, and then open the stopcock to fill the tip with solution and bring the level below the 0.00-mL mark. Read the lower level of the meniscus in the buret. Add 2 to 3 drops of phenolphthalein indicator, *not* universal indicator, to flasks 1 and 2, and place flask 1 under the buret tip and onto a white sheet. Lower the buret into the flask, open the stopcock with your left hand, and slowly swirl the flask with your right. After a short while, you will notice flashes of pink throughout the solution, which indicate the upcoming endpoint. Reduce the rate of flow of base as the pink color seems to remain longer. Close to the endpoint it is wise to add the base dropwise until the first indication of a *faint* pink color which persists for at least 30 s. The presence of a deep, dark red color indicates that you have added too much base, and your volume reading will be inaccurate.

　Record the exact volume of base used, and repeat for flask 2. Calculate the molarity *M* of the HBr for the two runs. If the values differ by more than 0.05 *M*, repeat a third time. Rinse the buret out with distilled H_2O when finished.

3. *Titration of Ba(OH)$_2$*. Using a graduated cylinder or pipet, remove two 30.0-mL portions of your Ba(OH)$_2$ from part I. Place each into a clean erlenmeyer flask labeled 1 and 2. Clean and fill a buret with *standardized HCl*, not 0.1 *M* HCl. Add phenolphthalein indicator, *not* universal indicator, to the flasks, and titrate until the red color changes to colorless. Use the same procedure and techniques described in part 2. Record all data. If the molarity (*M*) of the Ba(OH)$_2$ differs by more than 0.05 *M*, repeat a third time.

EXPERIMENT 19

ACIDS, BASES, AND SALTS
PRE-LABORATORY QUESTIONS

1. If 2 drops of universal indicator are added to a solution and the solution turns a yellow-green color, what is the pH of the solution?

2. If 3 drops of an H_2SO_4 solution neutralize 5 drops of an NaOH solution, what is the ratio M H_2SO_4/M NaOH?

3. Twenty-five milliliters of aqueous NH_3 solution required 50 mL of 0.48 M HCl. What is the molarity (M) of the NH_3 solution?

EXPERIMENT 19

ACIDS, BASES, AND SALTS
REPORT SHEET

Name _____   Section _____

I. SYNTHESIS

1. *Acids and bases from oxides*

   Distilled water pH = _____

   Al_2O_3 pH _____

   Write a chemical equation if a reaction occurred.

   P_2O_5 pH = _____

   Write a chemical equation if a reaction occurred.

   CaO pH = _____

   Write a chemical equation if a reaction occurred.

   SiO_2 pH = _____

   Write a chemical equation if a reaction occurred.

2. *Barium hydroxide*

   Write a balanced equation for the synthesis of $Ba(OH)_2$.

3. *Hydrobromic acid*

Write a balanced equation for synthesis of HBr.

4. *Synthesis of a salt*

Salt crystals and weighing paper   _____

Weighing paper   _____

Salt crystals   _____

Write a balanced equation for the reaction of $NiCO_3$ and H_2SO_4.

II. pH OF COMMERCIAL PRODUCTS

	pH	Acid or Base?	[H⁺]
Tap water			
Distilled water			
White vinegar			
Household ammonia			
Bowl cleaner			
Baking powder			
Sal-Soda			
Drāno			

III. GAS-PHASE ACID-BASE REACTION

1. Record your observations.

2. What substance is formed?

3. Is it soluble in water?

4. Write a balanced equation describing the reaction.

IV. ACID-BASE REACTIONS: TITRATIONS

1. *Naturally occurring indicators*

   Material used _____   Color in acid _____   Color in base _____

2. *Titration of HBr*

	Sample 1	Sample 2	Sample 3*
Final buret reading (mL)			
Initial buret reading (mL)			
Volume of NaOH used (mL)			
Volume of NaOH used (liters)			
M of NaOH used (see label)			
Moles of NaOH used			
Moles of HBr present			
M of HBr			

*If needed.

Average M of HBr _____

3. *Titration of Ba(OH)$_2$*

	Sample 1	Sample 2	Sample 3*
Final buret reading (mL)			
Initial buret reading (mL)			
Volume of HCl used (mL)			
Volume of HCl used (liters)			
M of HCl used (see label)			
Moles of HCl used			
Moles of Ba(OH)$_2$ present			
M of Ba(OH)$_2$			

*If needed.

Average M of Ba(OH)$_2$ _____

Problems

1. Suggest a method of preparation for:
   (a) HNO_3

   (b) $Mg(OH)_2$ ("milk of magnesia")

2. What effect (raise, lower, or no effect) would the following have on the pH of water?

   SO_2 _____ Na_2O _____ SnO_2 _____

3. If the solutions of $Ba(OH)_2$ and HBr that you prepared had been mixed together, what would be the result? Write an equation.

EXPERIMENT 20

OXIDATION AND REDUCTION

Materials Needed

Bi$_2$O$_3$

Bi lump

Fe lump

C

4-cm pieces Al wire

Crucible tongs

0.1 M KI

0.1 M KCl

Fresh chlorine water

0.05 M Br$_2$

0.1 M AlCl$_3$

0.1 M BiCl$_3$ in 2 M HCl

3 M HCl

Background

Consider the reaction of a metal with a nonmetal to give an ionic compound:

$$Ba + S \rightarrow BaS \quad \text{(contains Ba}^{2+}, S^{2-})$$

The metal, Ba, is *oxidized* because it loses two electrons: $Ba \rightarrow Ba^{2+} + 2e^-$. The nonmetal, S, is *reduced* because it gains two electrons: $S + 2e^- \rightarrow S^{2-}$. During an oxidation-reduction ("redox") reaction, there must be both a substance that loses electrons and a substance that gains them. Two related terms are *oxidizing agent:* a substance that causes oxidation of some other substance—and *reducing agent:* a substance that causes reduction. While it may seem confusing at first, during a "redox" reaction oxidizing agents get reduced and reducing agents get oxidized. In the above example, Ba is oxidized and is therefore the reducing agent; S is reduced and is therefore the oxidizing agent.

In a second example:

$$Fe_2O_3 + 2Al \rightarrow 2Fe + Al_2O_3$$

The Al is oxidized to Al^{3+} (in Al$_2$O$_3$) and is the reducing agent; Fe^{3+} (in Fe$_2$O$_3$) is reduced to Fe, and is the oxidizing agent. In a third example:

$$Cl_2 + 2KBr \rightarrow Br_2 + 2KCl$$

Convince yourself that Cl$_2$ is the oxidizing agent, Br$^-$ (in KBr) the reducing agent.

Industrially, oxidation-reduction reactions such as the second and third examples above are used to extract metals and nonmetals from minerals or seawater. In the case of metals, many occur as their oxides, and can be isolated by *reduction* with C or H_2 at high temperature:

Metal oxide $+$ C $\rightarrow$ metal $+$ CO_2 (CO may also form)

Metal oxide $+$ H_2 $\rightarrow$ metal $+$ H_2O(vapor)

Some specific commercial examples are:

Fe_2O_3 $+$ $3H_2$ $\rightarrow$ $2Fe$ $+$ $3H_2O$(vapor)

$2PbO$ $+$ C $\rightarrow$ $2Pb$ $+$ CO_2

The oxides of alkali metals, alkaline earth metals, lanthanides, and actinides cannot be reduced with C or H_2; these metals must be prepared by other means.

Some metals are oxidized by aqueous acids to give a metal ion and H_2:

Zn $+$ 2HCl(aq) $\rightarrow$ $ZnCl_2$(aq) $+$ H_2(gas)

Zn is the reducing agent, HCl the oxidizing agent. Actually, it is the hydronium ion (H_3O^+) present in aqueous HCl which is the true oxidizing agent. The Cl^- does not "do anything" except balance the charge of Zn^{2+}; Cl^- is therefore a "spectator ion." It is most important to recognize that $ZnCl_2$ is composed of Zn^{2+} and $2Cl^-$ ions, *not* Zn and Cl_2.

A reaction related to that of oxidation of metals by acids is the oxidation of a metal by the *ion* of a second metal. Usually these reactions take place in aqueous solution, but the second example above occurs in the dry state between solid Fe_2O_3 and solid, elemental Al.

The processes occurring here are:

Metal X $\rightarrow$ metal ion X^{n+} $+$ ne^{-1} (oxidation)

Metal ion Y^{m+} $+$ me^- $\rightarrow$ metal Y   (reduction)

As an example:

Mg $+$ Ni^{2+}(aq) $\rightarrow$ Mg^{2+}(aq) $+$ Ni

Mg is oxidized and Ni^{2+} reduced. This reaction occurs because Mg loses electrons easily, and Ni^{2+} accepts them.

Of course, the charges on metal ions are balanced by anions, which are not shown in the above example because they are "spectators" and do not participate in the reaction. For example, in the reaction

Mg $+$ $NiSO_4$(aq) $\rightarrow$ Ni $+$ $MgSO_4$(aq)

Mg is still the reducing agent and Ni^{2+} the oxidizing agent. SO_4^{2-} is a spectator ion. As another example:

$2Al$ $+$ $3Cu(NO_3)_2$(aq) $\rightarrow$ $3Cu$ $+$ $2Al(NO_3)_3$(aq)

Al is the reducing agent (and is oxidized); Cu^{2+} is the oxidizing agent (and is reduced). NO_3^- is the spectator ion.

Based on the ease of oxidation of metals in aqueous solution, an *activity series* may be constructed, in which metals easily oxidized are at the top of the series and those not easily oxidized are at the bottom. The deduction of an activity series can be best explained by use of a hypothetical series of experiments, with the indicated results.

1. $ZnSO_4$(aq) heated with Ag metal   no reaction

2. Ag_2SO_4(aq) heated with Cu metal   reaction to give Ag metal

3. H_2SO_4(aq) heated with Ag metal   no reaction

4. $CuSO_4$(aq) heated with Zn metal   reaction to give Cu metal

5. H_2SO_4(aq) heated with Zn metal   reaction to give H_2

6. H_2SO_4(aq) heated with Cu metal   no reaction

The conclusion to be drawn from each of these experiments is the following:

1. Ag is less active (less easily oxidized) than Zn, since Ag did not displace the Zn. Write Zn > Ag.
2. Cu is more active (more easily oxidized) than Ag, since Cu did displace the Ag. Write Cu > Ag.
3. Ag is less easily oxidized than H, since Ag did not displace H_2 from H_2SO_4. Write H > Ag.
4. Zn is more easily oxidized than Cu. Write Zn > Cu.
5. Zn is more easily oxidized than H. Write Zn > H.
6. Cu is less easily oxidized than H. Write H > Cu.

You now have Zn > Ag, Cu > Ag, H > Ag, Zn > Cu, Zn > H, and H > Cu. Note that SO_4^{2-} is a spectator ion and may be ignored.

Proceeding logically, note Zn > Ag, Zn > Cu, and Zn > H. That means Zn must be at the top of this particular activity series, and Zn is the most easily oxidized of these four elements. Next, note everything displaces Ag, and therefore Ag is at the bottom of the series. For the other two, H > Cu, so H is above Cu, and of course Cu > Ag:

ZN
H   activity series for
Cu   these four elements
Ag

Of course, a larger activity series could be constructed, in which you could determine, for example, that Mg is above Zn and Sn is below Zn but above H. Those elements *above* H on the activity series will react with aqueous acids, such as HCl or H_2SO_4, to give hydrogen gas (H_2) and the metal ion in solution. For example:

Sn + 2HCl(aq) → $SnCl_2$(aq) + H_2(gas)

On the other hand, Cu will *not* displace H_2 from HCl or H_2SO_4 because Cu is below H. Cu also will not displace Zn from $ZnSO_4$.

An activity series can also be constructed for nonmetals. In this case, the most easily *reduced* nonmetal is at the top. A partial list is:

F   activity series for
Br   some nonelements
S

From the list just given, you can deduce that

$$F_2 + Na_2S(aq) \rightarrow 2NaF(aq) + S$$

but

$$S + NaBr(aq) \rightarrow \text{no reaction}$$

Other types of "redox" reactions are known (not covered in this experiment) which involve molecules or ions reacting without the presence of elemental metals or nonmetals. Household "bleach" is an aqueous solution of sodium hypochlorite, NaOCl, containing the ions Na^+ and ClO^-. Some food stains react with bleach because they contain reducing agents, but not all stains are bleachable. Recognize that it is the ClO^-, not the Na^+, which is the oxidizing agent.

Procedure

I. REDUCTION OF A METAL OXIDE*

Weigh 9.0 g Bi_2O_3 and 0.7 g C into a 50-mL beaker, and mix with a glass rod. Pour the Bi_2O_3/C mixture into a crucible, cover, and heat strongly with a burner in the *hood* for 10 to 15 min. (Refer to Fig. 1 in Experiment 15.) By this time, red-hot molten bismuth metal (mp 271°) should be visible on removing the crucible lid with tongs. The metal is best removed by pouring it out while still molten onto a porcelain plate

Caution!! Do not attempt this without consulting the instructor.
Weigh the cooled chunk of Bi metal. This will be collected.

II. ACTIVITY SERIES FOR METALS

In the following procedure, the elements to be studied are aluminum (Al), bismuth (Bi), hydrogen (H), and iron (Fe).

Set up a boiling water bath by heating a large beaker of distilled water on a hot plate. Perform the following six experiments in test tubes, placed in the boiling water bath. Record your detailed observations of what occurs—or if no reaction occurs. A few bubbles of gas should not be considered a significant reaction.

1. 6 mL 0.1 M $AlCl_3$ + a lump or two of Fe metal. Does Fe replace Al? Look for the formation of gray metal powder. If no Al metal forms, you know Fe cannot replace Al, and Al > Fe in this activity series. If Fe *does* replace Al, then Fe > Al.
2. 6 mL 0.1 M $BiCl_3$ + one 4-cm piece of Al wire.
3. 6 mL 0.1 M $BiCl_3$ + a lump or two of Fe metal.
4. 6 mL 3 M HCl + one 4-cm piece of Al wire.
5. 6 mL 3 M HCl + a lump of Bi metal.
6. 6 mL 3 M HCl + a lump or two of Fe metal.

If in any of the above experiments no reaction seems to have occurred in the boiling water bath after 15 min, you may assume that particular experiment gave *no reaction*.

Metal residues will be collected. Do not dump them in the sink.

Based on the above experiments, deduce an activity series for these four elements, and explain your reasoning.

*A mixture of the less expensive but more toxic PbO (11.0 g) and C (0.5 g) may also be used.

III. ACTIVITY SERIES FOR NONMETALS

In the following procedure, the nonmetals to be studied are Br, Cl, and I.

Caution!! Elemental halogens are toxic and have unpleasant odors. Perform all experiments in the hood.

1. Pour 4 mL of 0.1 M KI into a test tube, and add 4 mL fresh chlorine water. Avoid an excess. Look for possible formation of iodine (I_2). Does Cl replace I? If not, then I > Cl.
2. Pour 4 mL 0.1 M KCl into a test tube, and add 4 mL 0.05 M Br_2. Result? Does Br replace Cl?
3. Pour 4 mL 0.1 M KI in a test tube, and add 4 mL 0.05 M Br_2. Result? Dispose of the reaction solutions in a container provided, *not* in the sink.

EXPERIMENT 20

OXIDATION AND REDUCTION
PRE-LABORATORY QUESTIONS

1. Based on the background of this experiment, predict the products of the following reactions; then balance them.

$H_2SO_4(aq) + Al \rightarrow$

$\rightarrow \Delta C + Fe_2O_3 \underset{\Delta}{\rightarrow}$

$Zn + AgNO_3(aq) \rightarrow$

2. Identify the oxidizing agent and reducing agent in each of the following:

$2BiCl_3(aq) + 3Zn \rightarrow 3ZnCl_2(aq) + 2Bi$

$H_2(g) + Ag_2O \underset{\Delta}{\rightarrow} 2\,Ag + H_2O(g)$

EXPERIMENT 20

OXIDATION AND REDUCTION
REPORT SHEET

Name —————————————— Section ——————————————

I. REDUCTION OF A METAL OXIDE

1. Weight of Bi metal obtained ——————— g.

2. Write a balanced equation for the reaction which occurred.

3. What is the oxidizing agent in this reaction?

II. ACTIVITY SERIES FOR METALS

1. In each of the six experiments, describe what occurred—or write "no reaction" if there was none.

2. Based on *your* data, deduce an activity series for these four elements. Explain your reasoning.

III. ACTIVITY SERIES FOR NONMETALS

1. In each of the three experiments, describe what occurred (or "no reaction").

2. Based on *your* data, deduce an activity series for the three nonmetals Br, Cl, I. Explain your reasoning.

EXPERIMENT 21

GLYCERIDES AND SOAPS

Materials Needed

10 g lard
10 g NaOH
30 mL ethyl alcohol
10 mL CH_2Cl_2
150 mL saturated NaCl solution
2 mL cottonseed oil
10 mL hexane
5% Br_2/H_2O in a dropper bottle
1 g commercial syndet

Universal indicator
1 mL motor oil
5 mL 1% $CaCl_2$
5 mL 1% $MgCl_2$
5 mL 1% $FeCl_2$
Boiling stones
Container for synthesized soap
Apparatus for reflux

Background

All of the materials that can be extracted into nonpolar solvents from animal or vegetable tissues are known as *lipids*. since many different types of organic molecules are soluble in the nonpolar solvents, a very large number of functional groups are found in the various molecules that are included in the category of "lipids." Most of the more common lipids, and therefore most of those studied by early investigators, have historically been classified on the basis of their physical properties: Oils are "greasy" liquids, fats are "greasy" solids, and waxes are solids that feel "waxy" to the touch. The historical classifications have continued to be used, but additional classes such as steroids and terpenes have been identified and included in the lipid category.

As more was learned about the structure of fats and oils, it became clear that they are chemically very similar: They are all *glycerides;* that is, they are triesters of glycerol (glycerin). In both fats and oils, the portions derived from carboxylic acids are generally unbranched and contain an even number of carbon atoms (most commonly 12 to 18).

For most glycerides, the differences in melting points, and hence the difference between fats and oils, are due to the amount of unsaturation in the groups indicated by R in the general formula. The presence of double bonds lowers the melting point. Oils are therefore often highly unsaturated, and fats are generally nearly saturated, but in other respects they are structurally very similar. Thus, fats and oils are best considered to be members of a single class of

$$
\begin{array}{cc}
\underset{\text{General structure}}{\underset{\text{of fats and oils}}{\begin{array}{c}
\text{O} \\
\| \\
\text{CH}_2\text{—O—C—R} \\
| \\
\text{O} \\
\| \\
\text{CH}_2\text{—O—C—R}' \\
| \\
\text{O} \\
\| \\
\text{CH}_2\text{—O—C—R}''
\end{array}}} &
\underset{\text{Glycerol}}{\underset{\text{(Glycerin)}}{\begin{array}{c}
\text{CH}_2\text{—OH} \\
| \\
\text{CH—OH} \\
| \\
\text{CH}_2\text{—OH}
\end{array}}}
\end{array}
$$

compounds that differ simply in their melting points. In fact, some people have stopped using the historical classification of fats and oils and simply use the terms "saturated fats" and "unsaturated fats." This causes some confusion (it probably would be better to refer to saturated and unsaturated glycerides), so it is necessary to remember that "unsaturated fats" are generally "oils" by the historical definition (which we will continue to use). Commercially, vegetable oils, such as cottonseed or peanut oil (unsaturated fats), can be converted into fats, such as margarine and solid "shortening" (saturated fats), by hydrogenating the double bonds.

Waxes are a much less homogeneous group of compounds than fats and oils. For instance, paraffin wax is a mixture of high-molecular-weight alkanes (in fact, alkanes are also known as "paraffins"). Carbowax is a synthetic polyether, and beeswax is an ester. Many waxes, like fats and oils, are esters, but, unlike fats and oils, waxes are not glycerides. Waxes are generally derived from monohydroxylic (one OH group) and dihydroxylic alcohols rather than from glycerol.

When it is treated with alkali, a fat or an oil is converted into glycerol and salts of carboxylic acids, known as *soaps*. In the past, this reaction has been the basis for the soap industry, and it is one of the sources of glycerol, which is used in antifreeze and in the manufacture of nitroglycerin. Although the sodium salts of long-chain fatty acids are water-soluble, the acids themselves are not. Thus, acidification of a soap solution causes the fatty acid to precipitate. Stearic acid ($C_{17}H_{35}COOH$) prepared this way is mixed with paraffin and used in the manufacture of candles.

$$
\underset{\text{A soap}}{R\!-\!\overset{\displaystyle \overset{\text{O}}{\|}}{C}\!-\!O^-Na^+} + H^+ \rightarrow \underset{\text{A fatty acid}}{R\!-\!\overset{\displaystyle \overset{\text{O}}{\|}}{C}\!-\!OH} + Na^+
$$

Because of problems with hard water, soaps have been replaced to a fairly large extent by synthetic detergents, or *syndets*. Strictly speaking, anything that has the ability to disperse oil and other water-insoluble organic materials in water, including soap, is a detergent, so that what we generally call "detergents" are synthetic detergents, and soap is a "natural" detergent. Both soaps and syndets contain anions that have a hydrophilic (water-seeking) negatively charged end and a long hydrophobic (water-repelling) hydrocarbon chain that is attracted to nonpolar substances such as oil and grease.

Both types of detergents disperse nonpolar materials in water by imbedding the hydrocarbon chain in the material. The negatively charged portion of the ion remains at the surface, where it can be attracted to the water, which allows the material to disperse. The ability of soaps and detergents to disperse nonpolar substances in water is the basis for their cleansing ability or detergency. In one portion of the experiment, we will compare the detergency of

$$R-\overset{\overset{O}{\parallel}}{C}-O^-Na^+ \qquad R-\underset{\overset{\parallel}{O}}{\overset{\overset{O}{\parallel}}{\underset{}{\bigcirc}}}-S-O^-Na^+ \qquad R-O-\underset{\overset{\parallel}{O}}{\overset{\overset{O}{\parallel}}{S}}-O^-Na^+$$

Sodium carboxylate　　　　Sodium aryl sulfonate　　　　　Sodium alkyl sulfate
(a *soap*)

Syndets

soaps and detergents in both soft (distilled) water and hard water (solutions containing calcium, magnesium, or iron salts).

Procedure

I. PREPARATION OF SOAP

Place 10 g of lard or other solid fat in a beaker and melt it on a steam bath. Dissolve 10 g of NaOH (*Caution: Corrosive!*) in 20 mL of distilled water. Pour the melted lard into a 100-mL round-bottom flask and add 20 mL of ethyl alcohol. Pour the NaOH solution into the flask, and assemble a reflux apparatus. Add several boiling stones, and reflux the mixture on a steam bath until a homogeneous solution is obtained (about 30 min). If material collects on the side of the flask, periodically swirl the flask to rinse the solid back into the reaction mixture. While you are waiting, go on to other parts of the experiment.

　　When the reaction is complete, pour the hot mixture (a towel may be used to hold the flask) into 150 mL of saturated NaCl solution contained in a 400-mL beaker. This should cause your soap to precipitate. Collect the soap on a Buchner funnel by suction, and wash it twice with 10 mL of distilled water. Continue to draw air through the soap until it is dry, and save it for part III.

II. PROPERTIES OF GLYCERIDES

1. *Solubility*. Place 10 drops (0.5 mL) of cottonseed oil in each of four test tubes.

   (a) To the first tube, add 1 mL of distilled water.

   (b) To the second, add 1 mL of ethyl alcohol.

   (c) To the third, add 1 mL of CH_2Cl_2.

   (d) To the fourth, add 1 mL of hexane.

   Shake each tube, and note whether or not the oil dissolves. If any of the solvents fail to dissolve the oil, add another 5 mL of solvent and again shake to see if the oil will dissolve. Record your observations.

2. *Saturation*.

   (a) Dissolve 5 drops of cottonseed oil in 1 mL of CH_2Cl_2 in a test tube. Add a solution of 5% Br_2/H_2O dropwise with shaking until the yellow color of the bromine is no longer rapidly discharged. Record the number of drops required.

(b) Repeat (a), using 5 drops of melted lard in place of the cottonseed oil. Again, record the number of drops of Br_2/H_2O required.

III. PROPERTIES OF SOAPS AND SYNDETS

Dissolve about 1 g of your soap in 50 mL of distilled water by heating on a steam bath. Similarly, prepare a solution of a syndet (1 g in 50 mL). These solutions are to be used for the remaining sections of the experiment.

1. *Alkalinity.* Test 1 mL of each solution with 1 drop of universal indicator. Estimate the pH of the solution.

2. *Detergent properties.*
   (a) *Soft water.* Place 4 drops of motor oil in each of three separate test tubes. Add 5 mL of distilled water to one tube, 5 mL of your soap solution to the second, and 5 mL of syndet solution to the third. Stopper the tubes and shake ach for about 30 s. Compare the abilities of the three liquids to disperse the oil.
   (b) *Hard water.*
      (i) Place 5 mL of your soap solution and 4 drops of motor oil in each of three test tubes. To the first, add 2 mL of 1% $CaCl_2$; to the second, add 2 mL of 1% $MgCl_2$; to the third, add 2 mL of 1 % $FeCl_3$. Stopper the tubes and shake them for about 30 seconds. Compare the abilities of the three liquids to disperse the oil.
      (ii) Repeat the procedure in (i), using 5 mL of syndet solution and 4 drops of motor oil in place of the soap and motor oil. Compare the "detergency" of soap versus syndet in hard water.

A container will be provided for collection of your unused soap.

EXPERIMENT 21

GLYCERIDES AND SOAPS
PRE-LABORATORY QUESTIONS

1. For glycerides, to what are the differences in melting points due?

2. Describe the cleaning action of a soap.

EXPERIMENT 21

GLYCERIDES AND SOAPS
REPORT SHEET

Name _____ Section _____

I. PREPARATION OF SOAP

1. The names of glycerides are based on the names of the acids from which they are derived. For example, triolein is derived from glycerol and 3 moles of oleic acid. Stearic acid is $CH_3—(CH_2)_{16}—\overset{\displaystyle O}{\overset{\displaystyle \|}{C}}—OH$. Draw tristearin.

2. Assuming that lard is composed completely of tristearin, write the equation for the preparation of your soap.

3. Give an appropriate chemical name for your soap.

4. What organic impurity was removed from the soap by washing with distilled water?

5. Why was ethyl alcohol added to the reaction mixture in the preparation of your soap?

II. PROPERTIES OF GLYCERIDES

1. Using the table on the next page, record your observations of the solubility of cottonseed oil in various solvents. The following abbreviations may be used:

   s = soluble   sl s = slightly soluble   i = insoluble

	1 mL of Solvent	6 mL of Solvent
Water		
Ethyl alcohol		
CH_2Cl_2		
Hexane		

2. How many drops of Br_2/H_2O solution were required for saturation?

   (a) Cottonseed oil   _____

   (b) Lard   _____

3. Which is the most unsaturated, cottonseed oil or lard?

Is this result expected?

III. PROPERTIES OF SOAPS AND DETERGENTS

1. *Alkalinity*. Record the pH of your soap solution and the syndet.

   (a) Soap solution.

   (b) Syndet solution.

   (c) How do you account for the acidity or basicity of your soap solution?

2. *Detergent properties*. Record your observations of the detergency of each of the liquids below.

   (a) Soft water
        (i) distilled water

        (ii) Soap solution

        (iii) Syndet solution

   (b) Hard water
        (i) With soap solution

             In $CaCl_2$ solution

             In $MgCl_2$ solution

             In $FeCl_3$ solution

        (ii) With syndet solution

             In $CaCl_2$ solution

             In $MgCl_2$ solution

             In $FeCl_3$ solution

Problems

1. Why was carbon tetrachloride at one time used for dry-cleaning? What dangerous disadvantage would there be if hexane were used instead?

2. Assuming your soap has the structure indicated in question 2 of part I, write an equation for the reaction of your soap with $FeCl_3$.

EXPERIMENT 22

IDENTIFICATION OF MINERALS

Materials Needed

Samples
Pan balance
Graduated cylinder
Steel file
Glass plate
Streak plate
Copper penny

Background

Minerals have definite chemical compositions and crystalline internal structures. Each mineral therefore possesses specific properties such as a crystal form, hardness, cleavage, specific gravity, color, luster, and streak. Cleavage may occur in one to six directions. An example is calcite, which has cleavage in three directions not at right angles. (See illustration on first page of lab report) Hardness is based on the Mohs scale of 1 to 10. Diamond is the hardest mineral, and it is assigned a number 10. Hardness is easily identified by mutual scratching of the mineral with a common article which has been ranked by its hardness. A steel file has a hardness of 6.5; glass, 5.5; knife blade, 5.1; copper penny, 3.1; and fingernail, 2.2. For example, calcite will scratch your fingernail, but it will never scratch a knife blade. Since its hardness is just less than that of a penny, it has a hardness of 3. Other attributes also have simple tests. See the instructor if help is needed using any of these tests.

Procedure

CAUTION: Do not break specimens without permission from the instructor.

Examine each mineral specimen provided for this exercise and briefly describe its physical properties as accurately as possible. Record each property systematically even though it may be completely obvious.

Table 22-1 Metallic luster characteristics

Streak		Type
Gray	Cubic cleavage, H = 2.5,* heavy sp gr = 7.6,† silver gray color	Galena (PbS)
Black	Magnetic, black to dark gray, H = 3, sp gr = 5.2, granular	Magnetite (Fe_3O_4)
Black	Steel gray; soft, marks paper, greasy feel, H = 1, sp gr = 2, dull	Graphite (C)
Greenish-black	Brass yellow, cubic crystals, H = 6 − 6.5, sp gr = 5, lacks cleavage	Pyrite (FeS_2)
Reddish-brown	Steel gray, black to dark brown-granular, fibrous or mica-like aggregates, crystals are thick plates, H = 5–6.5, sp gr = 5, lacks cleavage	Hematite (Fe_2O_3)

*H = hardness.
†sp gr = specific gravity.

Table 22-2 Nonmetallic luster and dark color

Harder than glass	Cleavage	Two directions at 90°, dark green to black, short 8-sided crystals, H = 6* sp gr = 3.5†	Pyroxene group: Ca, Mg, Fe Al
		Two directions at 60° and 120°, dark green to black or brown, 6-sided long prisms, H = 6, sp gr = 3–3.5	Amphibole group: Na, Ca, Mg, Fe
	No cleavage	Olive green glassy grains, conchoidal, clear, glassy luster, H = 6.5–7, sp gr = 3.5–4.5	Olivine: $(Fe, Mg)_2SiO_4$
		Red, brown, or yellow, glassy luster, conchoidal, 12-sided crystals, H = 7–6.5, sp gr = 3.5–4.5	Garnet: Fe, Mg, Ca, Al
Softer than glass	Cleavage	Brown to black, one plane cleavage, thin elastic sheets, H = 2.5–3, sp gr = 3–3.5	Biotite: K, Mg, Fe, Al
		Yellowish brown, resinous, 6-directional cleavage, yellowish brown or white streak, H = 3.5, sp gr = 4	Sphalerite: Zns
	No cleavage	Red, earthy, red streak, H = 1.5	Hematite: Fe_2O_3
		Yellow-brown streak, yellow-brown to dark brown, H = 1.5	Limonite: $Fe_2O_3 \cdot H_2O$

*H = hardness.
†sp gr = specific gravity.

Table 22-3   Nonmetallic luster-light color

Harder than glass	Cleavage	Good cleavage in two directions at 90°, pearly to vitreous, $H = 6–6.5*$ sp gr $= 2.5†$	Feldspar type—(1) potassium: pink white or green (2) plagioclase: white, blue-gray with striations
	No cleavage	Conchoidal, $H = 7$, vitreous, some clear, 6-sided prism terminated by 6-sided triangular faces	Quartz (S_1O_2): milky (white); smokey (gray to black), rose (light pink)
		Conchoidal, $H = 6–6.5$, translucent to opaque, dull or clouded luster	Agate: banded; flint: dark; chert: light; jasper: red; opal: waxy
Softer than glass	Cleavage	Perfect cubic cleavage, dissolves, colorless to white salty taste	Halite: NaCl
		Perfect cleavage in one direction, $H = 2$, white, transparent, sp gr $= 2.3$, may be an aggregate or fibrous	Gypsum: $CaSO_4 \cdot 2H_2O$
		Three directional cleavage effervesces in HCl, $H = 3$, colorless, white or pale yellow, sp gr $= 2.7$	Calcite: $CaCO_3$
		3-directional cleavage, effervesces when powdered, $H = 3.5–4$, white or pink crystals	Dolomite: $CaMg(CO_3)_2$
		4-directional cleavage, $H = 4$, sp gr $= 3$, cubic crystals, may be yellow, blue, green, or violet	Fluorite: CaF_2
		Planer cleavage, thin elastic sheets, $H = 2.3$, sp gr $= 2.8$, transparent and colorless in thin sheets	Muscovite: K, Al, H
		Green to white, soapy feel, pearly luster, $H = 1.2$, sp gr $= 2.8$, 1-directional cleavage, forms thin scales, may be foliated	Talc: Mg, H
	No cleavage	White to red, earthly soft, $H = 1.2$, becomes plastic when moistened, earthly odor	Kaolinite: Al, H

*H = hardness.
†sp gr = specific gravity.

1. Determine the type of luster, color, and streak.
2. Determine the type of crystal or texture of the crystal aggregates.
2. Test for hardness, cleavage, or fracture and other physical properties such as reaction with dilute HCl.

After recording these properties, try to identify the mineral by referring back to the data given in Tables 22-1 through 22-3, which list categories of mineral identification. The tables are arranged in a systematic manner on the basis of luster and color, hardness or streak, and cleavage or fracture. For example, suppose you observe that a mineral (1) is light-colored, (2) has a nonmetallic luster, (3) is softer than steel or glass but will scratch a penny, and (4) has good cleavage in four directions. You would then identify the mineral by referring to Table 22-3, nonmetallic luster, light color. You would next look under the list of minerals softer than glass and select the one which has a four-directional cleavage. The only mineral possessing these properties is fluorite.

EXPERIMENT 22

IDENTIFICATION OF MINERALS
PRE-LABORATORY QUESTIONS

1. What is a mineral? _____

   _____

2. Define:

   (a) Streak

   _____

   (b) Cleavage

   _____

   (c) Specific gravity

   _____

3. Why can we not simply classify minerals by their chemical composition?

   _____

4. What is meant by hardness? How could one show that a mineral was harder than calcite?

   _____

5. Rank the minerals you identified by hardness. Check each mineral by the method you described above and see if you are right.

   (a) Ranking

   _____

   (b) Checked?_____ (Yes or No)

   (c) Were there any inconsistencies? Explain:

   _____

6. Is ice a mineral? Why? _____

   _____

EXPERIMENT 22

IDENTIFICATION OF MINERALS
REPORT SHEET

Name _____   Section _____

Sample Number	Mineral Identified	Density	Crystalline Shape	Cleavage	Hardness	Color	Transparency	Luster	Streak

Cleavage in one direction.
Example: muscovite.

Cleavage in two directions at right angles.
Example: feldspar.

Cleavage in three directions not at right angles.
Example: calcite.

Examples of some types of cleavage. (*From "Physical Geology," W. Hamblin and J. Howard, Burgess Publishing Co., Minneapolis, 1975.*)

EXPERIMENT 23

LOCATING AN EARTHQUAKE

Materials Needed

Metric rule
Drawing compass
U.S. map (attached)

Background

One of the most frightening and destructive phenomena of nature is a severe earthquake. These unpredictable, catastrophic movements of the earth have caused a great number of human casualties and extensive property damage. Scientists have had some success in predicting earthquakes and in designing structures that will withstand the effects of earthquakes.

One of the early earthquakes for which we have detailed descriptive information occurred in 1755 near Lisbon, Portugal. Since that time detailed records have been kept. One of the early recorded earthquakes in the United States occurred in Madrid, Missouri, on December 16, 1811. The destruction of human life and property was slight because the most intense effects occurred in a sparsely populated region.

The 1906 San Francisco earthquake was one of the most destructive in the recorded history of North America. Other recent earthquakes include: Alaska, March 27, 1964; Mexico City, September 19, 1985; and San Francisco, October 17, 1989.

An earthquake is the oscillatory movement of the earth's surface that follows a release of energy in the earth's crust. In most cases a quake is caused by a dislocation of the crust, called a *fault* (see Fig. 1). Because of the stress within the crust the rocks break and generate vibrations called *seismic waves*. A recording of these waves is called a *seismogram*. The instrument used to record such waves is called a seismograph.

The epicenter of an earthquake is the point on the earth's surface directly above the center of the quake.

From the data expressed in the seismograms, the time, the epicenter, and the focal depth and its energy can be determined.

During a strong earthquake the vibrations are rather complex and require experienced interpreters. A seismogram of the 1989 San Francisco earthquake received at Albuquerque, New Mexico, is shown in Fig. 2. Less sensitive instruments give a less complicated readout.

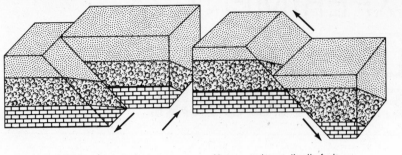

Movement along a strike-slip fault   Movement along a dip-slip fault   **FIGURE 1**

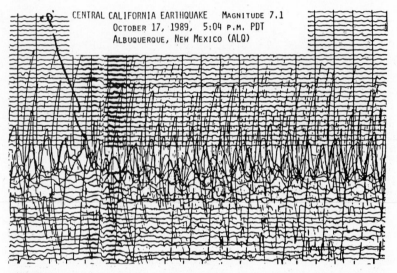

CENTRAL CALIFORNIA EARTHQUAKE   MAGNITUDE 7.1
OCTOBER 17, 1989, 5:04 P.M. PDT
ALBUQUERQUE, NEW MEXICO (ALQ)

FIGURE 2   (*Courtesy of the U.S. Department of the Interior, Geological Survey.*)

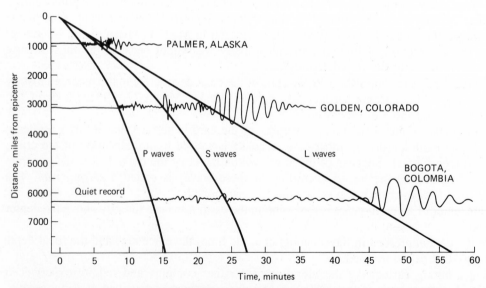

FIGURE 3   **Travel-time curves with idealized seismograms (earthquake records superimposed).** (*Courtesy of the U.S. Department of the Interior, Geological Survey.*)

The two general types of vibrations produced by earthquakes are *surface* and *body waves*. Surface waves travel along the earth's surface and probably cause most of the damage done by earthquakes. Body waves travel through the earth. Body waves are composed of two principal types: the P (primary) wave which compresses and dilates the rock as it travels forward through the earth, and the S (secondary) wave which shakes the rock sideways as it advances at approximately half the P-wave speed. The difference in speed between the two waves makes it possible for scientists to determine the epicenter of the quake. See Fig. 3. The amplitude of the wave is a measure of the energy of the quake.

Procedure

To understand how the difference in arrival time of two different waves traveling at different constant speeds can be used to determine the distance traveled, the following example may be helpful.

Consider two cars, one traveling at a constant speed of 40 miles per hour (mi/h), the other traveling at a constant speed of 60 mi/h. The distance to be traveled is 240 mi. The first car would require 6 h and the other 4 h, or a time difference of 2 h. It is obvious that if the distance were more than 240 mi, the time difference would be greater than 2 h. In this example the time difference of 2 h indicates that the distance traveled from the point of origin is 240 mi.

Figure 4 is a graph showing the time and the distance traveled by the P and S waves. The P wave travels 3000 mi in 8 min and the S wave travels 3000 mi in $14\frac{1}{2}$ min. In line with the discussion above, this difference in time of $6\frac{1}{2}$ min represents a distance of 3000 mi.

Assume that the difference in arrival time of the P and S waves at a certain seismograph station is 4 min. Find the position where the horizontal distance between the two waves is 4 min. Use a metric rule to measure this distance along the abscissa (horizontal). This distance is 20.0 mm. Slide the rule vertically upward, finding the position where the width between the two curves is 20.0 mm. Read the distance on the vertical (ordinate) that the difference in arrival time of 4 min represents. The approximate reading is 1600 mi.

This reading indicates that if the difference in arrival time of the two waves is 4 min, the earthquake occurred at a distance of 1600 mi from this station.

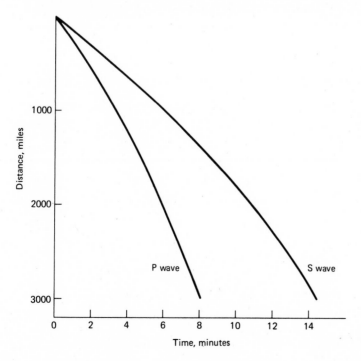

FIGURE 4   Travel-time curves.

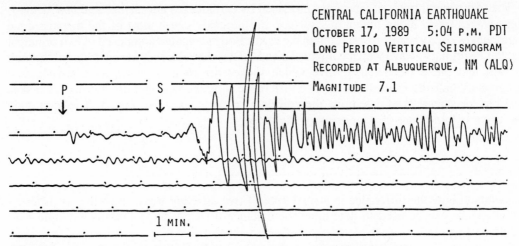

FIGURE 5 (*Courtesy of the U.S. Department of the Interior, Geological Survey.*)

If a circle is drawn about this station using a scale representing 1600 miles, it is obvious that the epicenter is located somewhere on this circle. Since this is very indefinite, it is necessary that information be received by other stations.

If information from another station is received, a second circle could be drawn. The location of the epicenter would be limited to the two positions where the two circles intersect. The introduction of information from a third reporting station would limit the number of intersections to one, thus determining the exact location of the epicenter.

Note that in using this method of locating the epicenter, it is necessary to know the arrival time of both the S and P waves at each of at least three stations.

Figure 5 is a seismogram recorded at Albuquerque, New Mexico, in October 1989. Note the dots are 1 min apart. The time between the arrivals of the P and S waves can be determined

FIGURE 6 The United States. (Scale: 1 cm = 150 miles.)

from the seismogram. Using the curves discussed previously, the distance to the epicenter is readily determined.

Determine the location of a quake's epicenter by using the data received from three seismograph stations. The data are simulated and do not represent an actual earthquake.

The stations reporting and the data received are as follows:

Spokane, Washington:	Arrival of P waves 10:29:24
	Arrival of S waves 10:32:12
Lincoln, Nebraska:	Arrival of P waves 10:30:10
	Arrival of S waves 10:33:40
Louisville, Kentucky:	Arrival of P waves 10:31:51
	Arrival of S waves 10:36:33

Record these values in the table given on the Report Sheet. Calculate the difference in arrival time for each station and record these values in the table. From the difference in arrival times of the P and S waves at each station determine and record the distance to the epicenter from that station. (Use Fig. 4.)

The map of part of the United States (Fig. 6) is drawn to a scale of 1 cm = 150 mi.

Draw a circle about each reporting station using the determined distance to the epicenter using the scale on the map—i.e., for a distance of 900 mi the radius of the circle would be 900 ÷ 150, or 6 cm. Note after drawing the first circle that all that is known is that the epicenter is someplace on this circle. The second circle should intersect the first in two places. The third circle should intersect the other two in only one of these two places, thus locating the epicenter.

EXPERIMENT 23

LOCATING AN EARTHQUAKE
REPORT SHEET

Station	Arrival		Arrival Time Differences	Distance	Radius
	P waves	S waves			

Problems

1. Where did the earthquake occur? _____

2. Is this location a suspected earthquake area? _____

3. What time did the earthquake occur? _____

EXPERIMENT 24

LENGTH OF DAY AND NIGHT

Materials Needed

Globe (approximately 8 in)
Light source—100-W frosted bulb mounted on a porcelain base or reflector-type desk light

Background

As you know the sun rises earlier in the summer than in the winter and sets later in the summer than in the winter. Table 24-1 gives the times of sunrise and sunset for several latitudes.

Table 24-1

| | Latitude | | | | | |
| | 0° N | | 30° N | | 60° N | |
	Sunrise, A.M.	Sunset, P.M.	Sunrise, A.M.	Sunset P.M.	Sunrise A.M.	Sunset P.M.
Jan. 1	6:00	6:08	6:56	5:12	9:02	3:06
Feb. 1	6:10	6:18	6:51	5:37	8:15	4:13
Mar. 1	6:09	6:16	6:28	5:58	6:55	5:30
Apr 1	6:00	6:07	5:49	6:19	5:24	6:45
May 1	5:53	6:00	5:17	6:37	3:56	7:59
June 1	5:54	6:01	5:00	6:56	2:48	9:08
July 1	6:00	6:08	5:02	7:05	2:40	9:26
Aug. 1	6:02	6:10	5:18	6:55	3:36	8:34
Sept. 1	5:56	6:03	5:44	6:22	4:50	7:10
Oct. 1	5:46	5:53	5:53	5:45	6:03	5:36
Nov. 1	5:40	5:48	6:13	5:14	7:24	4:04
Dec. 1	5:45	5:54	6:38	5:00	8:35	3:03

Procedure

Graphs will be drawn to show this information. From the graphs one may note the length of daylight during the different months and that this length varies at different latitudes. Houston, Texas, has a latitude of approximately 30°N. The first graph will be a line graph showing sunrise and sunset for 30°N latitude. Label each curve as sunrise and sunset for 30°N latitude. Lightly darken the part of the graph which indicates darkness. Your graph should clearly indicate the variation in length of day and night.

Repeat for the other latitudes (0° and 60°) shown in Table 24-1. In drawing the curves, use a separate color for each latitude (same color for both sunrise and sunset). To the right of each curve, letter the latitude of the curve.

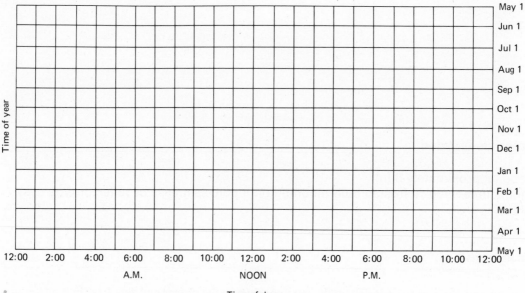

EXPERIMENT 24

LENGTH OF DAY AND NIGHT
REPORT SHEET

1. From your curves, about what date does Houston (use 30°N latitude) have the longest period of daylight? _____

2. How many hours? _____

3. At about what date does Houston have the shortest period of daylight? _____

4. How many hours? _____

5. Fill in the following table:

Latitude	0°	30° N	60° N
Longest day (hours of daylight)			
Date of longest daylight			
Shortest day (hours of daylight)			
Date of shortest day			

6. Which of the latitudes has the longest daylight? _____

7. Shortest? _____

8. Which latitude has the least variations of length of daylight? _____

9. From the graph, when does the sun rise at 6:30 in Houston? (30°) _____
   (*Note*: two dates)

10. When are there 12 h of daylight in Houston? (These dates are called equinoxes.) _____

    _____ (*Note*: two dates)

Note that the length of daylight varies very little at the equator (0° latitude) for the different times of the year while at places far from the equator the length of the day and night varies considerably.

Place a globe so that it is inclined toward (June 21 position) a strong source of light. (Fig. 1). Stand to one side so that a line between you and the globe is at right angles to a line between the light and the globe. Have one of the members of your group rotate the globe.

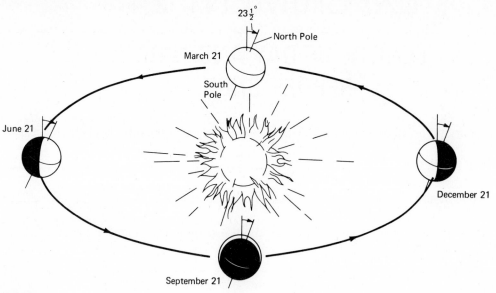

FIGURE 1

Which region has 24 h of daylight? _____

Which region has 24 h of darkness? _____

Note the direction at which the sun's rays strike the earth's surface in both the northern and southern hemisphere.

Which section is having summer? _____

Winter? _____

In respect to the equator where will the sun appear to be at 12:00 (noon)? _____

Complete the following table for the June 21 position from your observations above:

Length of daylight

Latitude	Date			
	Mar. 21	**June 21**	**Sept. 21**	**Dec. 21**
0° (Equator)				
23½° (Tropic)				
66½° (Arctic Circle)				
90° (North Pole)				
South Pole				

Move the globe around the light 90° in a counterclockwise direction (looking down from the top). Keep the axis inclined in the same direction as for the June 21 position. This is the September 21 position, or autumnal equinox. Again place yourself at a right-angle position to the line joining the light and the globe and in line with the earth. Complete the above table for this position. Note the angle at which the sun's rays strike the earth.

What season are we having in the northern hemisphere? _____

At this time to an observer in the northern hemisphere, would the sun appear to be moving

north or south? _____

Why is September 21 called the *autumnal* equinox? _____

With respect to the equator where will the sun appear to be at 12:00 noon? _____

Move the globe another 90° in the same direction, keeping the axis pointing in the same direction. This will represent the December 21 position, or winter solstice. Again take a position at right angles to a line joining the light and globe.

As one rotates the globe, which region now has 24 h of darkness? _____

Why is December 21 called the winter solstice? _____

Complete the table for this position. Note the angle at which the sun's rays are striking the earth's surface.

What season are we having in the northern hemisphere? _____

In the southern hemisphere? _____

At this position the earth is nearest the sun. Do our seasons depend upon the distance

from the sun or the inclination of the earth's axis? _____

If the distance factor were the only thing to be considered (that is, if the earth's axis were

not inclined), what would happen to our seasons? _____

At this position, the sun is its greatest distance south of the equator. Since the earth's

axis is inclined $23\frac{1}{2}°$, how far south does the sun appear to be from the equator? _____

_____

What do we call the imaginary line drawn around the globe at this point? _____

A similar line is found in the northern hemisphere when the sun is farthest north. What

is this line called? _____

Again move the globe 90° as before, keeping the axis inclined to the same direction. Assume the same relative position for observing the globe. The position of the globe represents the March 21 position, or vernal equinox. Complete the table for this position. Note this is the first day of spring position.

Why is the climate moderate at this time? _____

Which way does the sun appear to be moving? _____

If the earth's axis were perpendicular to the plane of its orbit, which effect would that

have on the length of the day? _____

On seasons? _____

If the axis were inclined more than $23\frac{1}{2}°$, how would this affect the variation in the length

of the day? _____

EXPERIMENT 25

EARLY MEASUREMENTS IN ASTRONOMY

Materials Needed

Part 1:
   Attached highway map with latitude and distance scale metric rule
Part 2:
   4 in × 6 in card with pinhole
   4 in × 6 in blank card with 8-mm circle
   Meter stick
   Meter stick card supports

PART 1. CIRCUMFERENCE OF THE EARTH

Background

Early observers of the heavens made many observations of the earth, moon, sun, and other stars. Records of these observations were kept in the local libraries. The Greek scholar Eratosthenes was the librarian at Alexandria and was acquainted with these records. Early Greek astronomers were interested in measuring the distances and sizes of the sun, moon, and earth and their distances apart. The first measurement to be made was the size of the earth itself. Using available records, Eratosthenes was the first to develop a means for measuring the circumference of the earth. Details of this information are shown on the following page. Note that to make such a measurement one needs to know the angle and the distance between the two stations involved. The two stations must have the same longitude [north (N) or south (S) of each other]. The following figures illustrate approaches to obtain this information. Note that the methods above could be used to determine the angle between the two stations.

Procedure

It is possible for you to measure the latitude of two stations north and south of each other and some distance apart. We will use a convenient method. On page 231 is part of a highway map for a section of southeast Texas that you can use to measure both the latitude and the distance

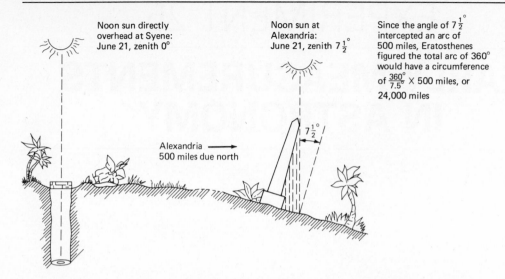

Noon sun directly overhead at Syene: June 21, zenith 0°

Noon sun at Alexandria: June 21, zenith $7\frac{1}{2}°$

Alexandria →
500 miles due north

$7\frac{1}{2}°$

Since the angle of $7\frac{1}{2}°$ intercepted an arc of 500 miles, Eratosthenes figured the total arc of 360° would have a circumference of $\frac{360°}{7.5°} \times 500$ miles, or 24,000 miles

between the two stations. To find the latitude, measure the distance between two lines of labeled latitude (assume 27° and 28° with a distance of 45 mm between them). Now measure the distance from the 27° to the station (assume this measurement to be 37.5 mm). This means that the station is $\frac{37.5}{45}$ of a degree, or 0.83 above 27°. The latitude would be 27.83° N.

You use the same method for determining the latitude of a second station. To determine the distance between the two stations, use the scale appearing at the lower right.

It is suggested that you use the intersection of highways 77 and 83 (west of Harlingen, just above Brownsville) for station 1, and Lockhart (south of Austin on 183) as station 2.

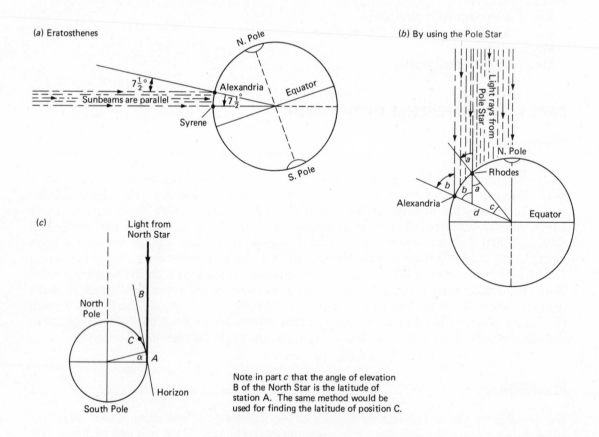

(a) Eratosthenes

$7\frac{1}{2}°$
Sunbeams are parallel
N. Pole
Alexandria
$7\frac{1}{2}°$
Equator
Syrene
S. Pole

(b) By using the Pole Star

Light rays from Pole Star
N. Pole
Rhodes
a
b
b a
Alexandria
d
c
Equator

(c)

Light from North Star
B
North Pole
C
α A
Horizon
South Pole

Note in part c that the angle of elevation B of the North Star is the latitude of station A. The same method would be used for finding the latitude of position C.

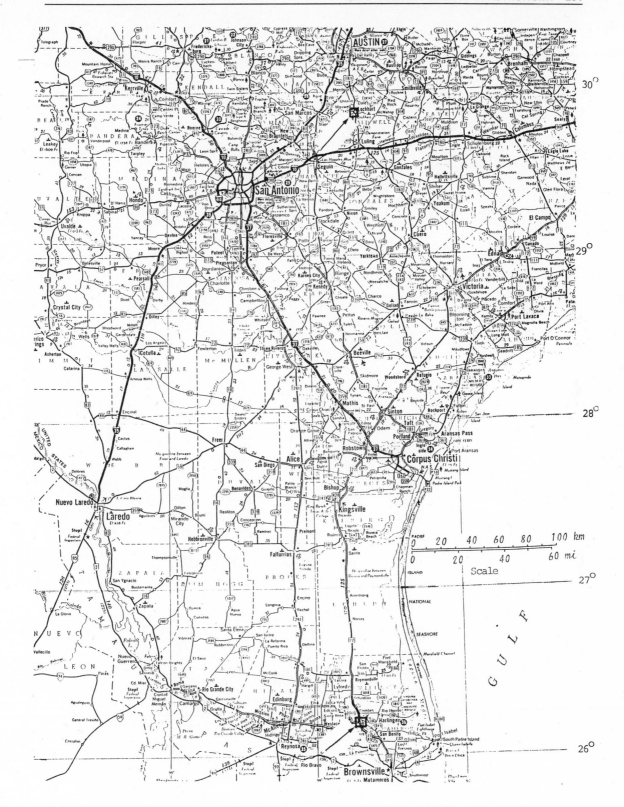

EXPERIMENT 25

EARLY MEASUREMENTS IN ASTRONOMY, PART 1
REPORT SHEET

Name _____   Section _____

	Station 1	Station 2
	_____	_____
Latitude of	_____	_____

Difference between latitudes, θ degrees   _____

Distance between Station 1 and Station 2 (X)   _____

Using the above data, calculate the circumference of the earth. Compare your answer with the accepted value:

$$\frac{X}{\theta} = \left(\frac{C}{360°}\right)°$$ _____ = _____ $C =$ _____

Accepted value _____

Percent of difference _____

Show your measurements and calculations for determining the latitude of each station and the distance between the two.

You may wish to contact someone that lives several miles north or south of you to make altitude observations of the North Star. One could use these date to repeat the above experiment.

PART 2. MEASURING THE DIAMETER OF THE SUN

Background

Light reflected through a pinhole will form images as shown in the figure below. The rays of light form triangles that are similar and the relationship $S_i/D_i = S_o/D_o$ is true.

    If you use light coming through a window from the outside you can move the image card and obtain many images of the outside scene. During an eclipse of the sun, tree leaves form many pinholes and you will see many images on the ground.

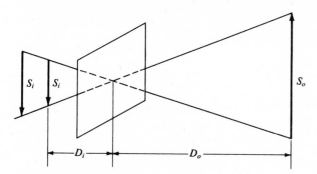

Procedure

Assemble the meter stick and card as shown in the figure below. Place the cards on the meter stick with the pinhole card A at the 90-cm mark. Aim the 100-cm end of the meter stick at the sun. (***Caution: Do not look directly at the sun.***) Move the image card B until the sun's image just covers the circle. (Circle approximately 8 mm in diameter.) When the bright image just fills the circle, measure the distance between the pinhole card A and the image card B.

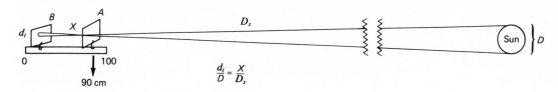

EXPERIMENT 25

EARLY MEASUREMENTS IN ASTRONOMY, PART 2
REPORT SHEET

Name _____   Section _____

1. Record the distance D_s from the earth to the sun using the accepted value.
2. Record the image distance X in centimeters.
3. Record the size of the sun's image in centimeters (d_i).
4. Calculate the sun's diameter D using the proportion

$$\frac{d_i}{D} = \frac{X}{D_s} \qquad D = \frac{d_i(D_s)}{X} = \underline{\hspace{2cm}} = \underline{\hspace{3cm}}$$

5. Find the % of error between your value and the accepted value.

   (a) Distance from earth to sun (D_s)   _____

   (b) Distance between card A and B (X)   _____ cm

   (c) Size of sun's image (d_i)   _____ cm

   (d) Sun's diameter (D)   _____

   (e) Accepted value for sun's diameter   _____ mi

   (f) % error   _____

EXPERIMENT 26

TOPOGRAPHY
OF THE MOON

Materials Needed

Moon photographs (included in the experiment)
Metric rule

PART 1. INTERPRETATION OF MOON PHOTOS

Background

The moon has long been one of our most fascinating neighbors. Because of its nearness, it appears as one of the largest and brightest bodies in the sky.

Prior to the mid 1960s scientists were able to observe the moon through telescopes and to photograph its surface only from the earth. Light-covered areas were interpreted as highlands and dark areas as lava filled basins. Craters were interpreted to be of both volcanic and impact origin.

In the mid-1960s NASA conducted two programs involving the moon. The first was a lunar orbiter. The lunar orbiter provided many excellent photographs of the moon's surface. The second mission, Surveyor, successfully landed five spacecraft on the moon to collect data regarding the chemical and physical properties of the surface materials. The Apollo program to send humans to the moon for direct exploration and to return with moon rock samples proved successful. The rock samples proved to be similar to earth materials.

We see the moon by reflected light. Although it appears to be very bright, it reflects only about 10 percent of the light striking its surface.

With a good pair of binoculars for viewing the half moon, one can see many of the larger craters, especially near the sunrise or sunset line (terminator). Streaks (see the first photo, Fig. I, sections A3 and A4 and B3 and B4) called *rays,* can best be observed at full moon. For most viewing, the period of greater shadows is preferable.

The surface of the moon is dominated by two kinds of landscape: dark lava plains with few craters known for historical reasons as *seas* or *maria* (Fig. I, sections D2 and D3 and E2 and E3), and lighter-colored, heavily cratered, highlands (Fig. I, sections G1 and G4).

Theories regarding the creation of craters include: volcanic activity, meteorites, and gas bubbling from the interior. Crater sizes (Fig. I, sections A3 and A4 and B3 and B4) vary a

239

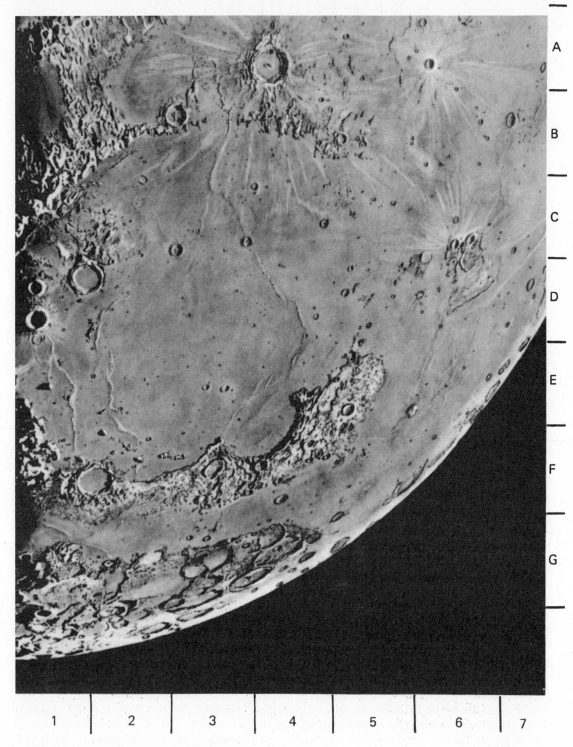

FIGURE I (*Courtesy of Lick Observatory.*)

FIGURE II   (*Courtesy of Lick Observatory.*)

great deal, from pin size to 150 mi in diameter and 20,000 ft in depth. The length and darkness of the shadow within the crater give an idea of the depth and steepness of the wall. The position of the shadow gives the direction of the sun. See Fig. I, section D1. The sun's rays are coming from the right.

The period of rotation of the moon is approximately 4 weeks. Such long periods of daylight and darkness result in extreme temperature changes. During the period of darkness, the temperature drops to $-162°C$; during the long period of sunshine, it rises to $120°C$.

Because of the moon's small size, its gravity is about one-sixth that of the earth. This is not enough to retain any but the heavier gases and vapors. Without an atmosphere, the weathering processes that have so drastically modified the earth's surface do not act on the moon. The moon is without winds, weather, or forces of erosion other than the effects of extreme heat and cold and bombardment by meteors. The number of meteorites striking the moon's surface is large since there is no atmosphere to burn any of them up.

Very prominent, rather smooth, dark, usually depressed plains are seen on the moon's surface. These areas are called *maria* (singular *mare*). About one-half of the moon's visible surface but little of the backside is covered with maria (Fig. I, sections B2 and B3 and C2 and C3). Circular maria (Fig. I, sections B2 and B3 and C2 and C3) are surrounded by rims or ring anticlines beyond which lie the plains. The dark regions seem to have been caused early in the moon's history by the impact of large bodies and subsequent flooding by lava from the moon's interior.

The valley-like depressions on the moon's surface having steep sides, flat bottoms, and varying widths, are called *rills* (see Fig. II, sections D4 and E4). Some rills extend for hundreds of miles. The rills appear dark because the sun's rays do not always penetrate to the bottom. They are generally thought to be features bounded by faults and formed by subsidence of the surface.

In contrast to the rills, *mare ridges,* or *wrinkle ridges,* rise above (Fig. I, sections A7 and B7) the surface of the moon. These are found only on maria and are best seen when sun rays slant across the moon at a low angle. They average less than 1000 ft in height and perhaps 10 to 20 mi in width. Wrinkle ridges are probably caused by compression, in contrast to rills which probably are due to tension. Lava flow in the maria caused crustal bending and sagging of the moon's surface. This movement resulted in tension cracks or rills at the edges of the maria. Compression, or wrinkle ridges, may have formed in the central part of the maria during movement or cooling and contraction of the moon's surface.

Other distinctive features of the moon's surface are the rays or bright streaks that radiate from craters. Some of the rays extend as far as 2000 kilometers (km). The rays probably consist of pulverized rock blasted from the moon's surface during the formation of a crater. The rays may darken during bombardment by cosmic material or fade when exposed to solar radiation. The darkening process may make it impossible to recognize the rays. For this reason, rays may be a clue to the relative age of the craters. Tycho, with its spectacular ray system, is considered to be one of the younger craters.

Small pits in the moon's surface are called *chain craters* (Fig. II, section F2). These are small in diameter and are aligned in chains and may extend more than 100 mi. Chain craters are found in the maria. However, a few extend into the higher, light-colored surrounding area. Chain craters probably resulted from violent eruptions of gas (volcanism) which threw out dust and fragments of surface material or from the drainage of surface material into voids created by faults.

Procedure

This exercise, interpreting moon photos, can be done as a group or as an individual activity using the questions on the Report Sheet as topics for discussion.

EXPERIMENT 26

TOPOGRAPHY OF THE MOON, PART I
REPORT SHEET

Name_____   Section_____

I. FIGURE I

Copernicus is the large crater (see section A4). What is its diameter, if the scale is 1 cm

= 110 km? _____

How far out do the rays extend? _____

What do most scientists believe concerning the origin of these rays? _____

What is the center of Copernicus, a peak or a crater? _____

Near the bottom there are several large craters (see section G3). Is this area young or

mature? _____

Is the crater in sections F1 and F2 shallow or deep? _____

Note the circular maria in sections E3 and E4 and F3 and F4. Is there evidence of lava

flow in this region? _____

One may note several lines vertically (sections B3 and B4 and C3 and C4). Are these

valleys or mountain ridges? _____ How do you know? _____

What are these called? _____

What do you call the dark smooth surface in the central portion of the photo (sections

D3 and E3)? _____

How do you account for some of the craters appearing elliptical in shape, while others

appear circular (sections F1 and F2)? _____

How long is the rill in section A3? _____

The Apollo 15 mission landed just below and to the left of Copernicus in the Apennine

Mountain area (sections B2 and B3). Why do you think they chose this area? _____

Do any of the craters have central crater peaks (see sections B2 and B3)? _____

Two rather small distinct craters are located on the extreme left-hand side (section D1).

Are these craters shallow or deep? _____   Why do you say this? _____

Is there evidence of small mountains just below these craters (see sections E1 and E2 and

F1 and F2)? _____

II. FIGURE II

In the lower part of the picture (section F4), there are two medium-sized craters. Are
these young or mature? Compare the depth and the steepness of their walls.

How do you account for the very rough surface in the lower part of the picture (section
G5)?

Note the line running from the center down and left in sections C5 to F1. Is this a ridge
or a valley? What caused it? What is it called? Compare the steepness and depth at various
points.

Is the large crater on the left (section D1) deep or shallow? Do you see a dome within?

A medium-sized crater in section E6 has a bright ring all around it. How do you account
for this? Is this crater deep or shallow? Is the right-hand side steep or sloping? What do
you see in the middle?

Note the rill chain in sections E2 and F2. Note the very small and young crater in section
D6. The rays are almost star-like. The geologists at NASA have made a statistical study
of the moon's surface. One may assume that the chance of a very large crater occurring
is less than that for smaller craters. Based on this assumption, how would you compare
the age of the left-hand side with that of the upper center in the photo?

PART 2. DEPTH AND DIAMETER OF A CRATER ON THE MOON

Background

Since early times people have been interested in the measurement of the moon's surface features. Interest in the depth and diameter of various craters has been a part of this curiosity.

To determine such measurements, the crater located in sections D1 and E1 of Fig. I has distinct features making required measurements possible. For these reasons this crater has been selected for this study.

The measurements required to determine the depth include the length of shadow, distance from the terminator, and the known radius of the moon.

To develop the math required refer to Figs. III and IV. The moon is a sphere, and the lines shown do not lay in the same plane since the radius line would slant toward the center of the moon. This would complicate the geometry. By rotating the moon about CY, the crater can be changed from position P' to P, as shown in Figs. III and IV. Since all distances to be measured are now in the same plane, the equation $D/S = AP/R$ can be used to determine the depth D.

The floor (GEF) of the crater is level with the moon's surface. If D = depth, S = length of shadow, AP = distance to terminator, and R = radius of moon, then (using Fig. IV)

$$AP \perp AC \qquad EF \perp CP$$

Therefore

$$\angle ACP = \angle FEP$$

and ACP and FEP are right triangles. Therefore, triangles ACP and FEP are similar and

$$\frac{D}{S} = \frac{AP}{R}$$

from similar triangles

$$D = \frac{S\,(AP)}{R}$$

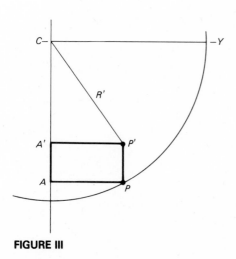

FIGURE III

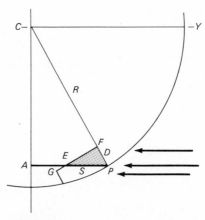

FIGURE IV

The above proportion is developed only for your information. Your understanding of its derivation is not necessary for it to be a part of the solution.

Procedure

Using the moon photo in Fig. I, measure the scalar distance (distance on the moon photo in centimeters) for the moon's radius, the scalar distance AP from P to the terminator, and the length S of the shadow. The accepted diameter of the moon is approximately 3500 km.

EXPERIMENT 26

TOPOGRAPHY OF THE MOON, PART 2
REPORT SHEET

Name _____   Section _____

    Scalar radius R of moon _____ cm

    Scalar distance AP to terminator _____ cm

    Length S of shadow _____ cm

1. Using the above values and the equation $D = S(AP)/R$ determine the scalar depth D for the crater. Show your work in the space below.

2. Scalar depth of the crater is _____ cm. Using the given diameter of the moon and the measured scalar radius of the moon's in Fig. I, determine the scale used in the photograph. From this value and the scalar depth above determine the real depth of the crater.

    (a) Diameter of moon _____ km

    (b) Radius of moon _____ cm

    (c) Scalar radius _____ cm

    (d) Scale on photograph, 1 cm = _____ km

    (e) Scalar depth _____ cm

    (f) Scale on moon's photo, 1 cm = _____ km

    (g) Real depth crater _____ km

(*Note:* This depth will be from the top of the crater's rim.)

3. To determine the diameter of the crater, measure its scalar diameter. Using the scale on the photo calculate the real diameter of the crater.

    (a) Scalar diameter _____ cm

    (b) Real diameter _____ km

4. Since you do not have known values for the depth and diameter of this crater check references for what astronomers believe to be a range of values. State your findings below.

EXPERIMENT 27

SPECTROSCOPY

Materials Needed

Hand spectroscope
Wood grating spectrometer (Sargent Welch)
Light sources
Power supply for light sources: hydrogen, mercury, sodium
Incandescent light bulb
Fluorescent light fixture

Background

This experiment will introduce you to the subject of spectroscopy. You will use some of the equipment used by physicists and astronomers to examine visible light. Different light sources will be observed to illustrate the types of spectra and how they provide clues to the composition of stars. From the spectral appearance of a star, its spectral class is determined. This gives direct information on the temperature of the star. From this information and the apparent brightness of the star, its distance can be determined. There are three major types of spectra: continuous, absorption, and emission. A continuous spectrum is rainbow-like and its colors smoothly grade from violet to red with no dark or bright lines apparent. This type of spectrum arises usually from heated solids (incandescent sources) but may also be produced by dense hot gases such as that below the sun's surface (photosphere). Emission and absorption spectra arise from low-density gases. When such a gas lies between an incandescent source and the observer, certain discrete lines or wavelengths of light are absorbed. The pattern of dark lines is a characteristic of the type of gas present. Thus, the resulting spectrum has the appearance of a continuous spectrum with dark lines superimposed. This type of spectrum is referred to as an *absorption* or *dark line spectrum*. The sun emits this type of spectrum. Can you guess why? When the gas absorbs light, it will re-emit the same discrete wavelengths in the form of emission or bright lines. Thus the glowing gases in the discharge tubes viewed in this lab give off a pattern of emission lines viewed against a dark background. This pattern is called an *emission* or *bright line spectrum*.

Procedure

1. On your desk is a tube about a foot long. This is a simple hand-held spectroscope. You look through the end which has the round opening. Look toward the side of the tube, not straight ahead.
   (a) Go to the window and look at the sky. Describe the color pattern you see (use your Report Sheet for the description). This is due to the scattered light of the sun. If you had a better spectroscope you would see dark lines superimposed on the pattern.
   (b) Look at the incandescent light bulb and describe its spectra.
2. In this step you will use a more complicated device, the Wood spectrometer. This one spreads out the pattern on a scale which can be used to measure wavelength. This instrument uses a diffraction grating. Other spectroscopes may use a prism to split up the light.
   (a) Some of the light sources use high voltage. Avoid touching bare wires. Your instructor will connect the special power supplies to the light sources. You will plug the standard plugs into the 117-V alternating current (ac) socket on the table's power supply since only one can be connected at a time.
   (b) Find the metal slit on the side of the spectroscope. A light source is positioned so that it shines through the slit.
   (c) Look through the Wood spectrometer at the hydrogen source (stars are predominantly hydrogen and helium by composition). The wavelength scale on which the spectral lines are projected is in nanometers (nm) (1 nm $= 10^{-9}$ m). Visible light ranges in wavelength from about 400 nm (violet) to 700 nm (red). Record the position of each major line viewed and note its color. Copy the spectra by drawing all major lines in their approximate position in the boxes shown in the Report Sheet. Write the wavelength read from the scale on the side of each drawn line.
   (d) Repeat steps (b) and (c) for the mercury and the sodium source.
   (e) What *type* of spectra is given off from these sources?
3. Compare your measured wavelength values to the actual wavelength of the red line (H_α) of hydrogen (actual wavelength 486.1 nm), the aqua line (H_β) of hydrogen (actual wavelength 656.3 nm) and the yellow sodium doublet (average wavelength 589.3 nm). Compute a percentage error for each.
4. Observe the flourescent light source with the Wood spectrometer. What two types of spectra are seen? Why?
5. Briefly explain the Bohr model of the hydrogen atom (see textbook). How are absorption lines explained? The emission lines?

EXPERIMENT 27

SPECTROSCOPY
REPORT SHEET

Name _____   Section _____

From Procedure:

1. Type of spectra seen:

   (a) Sky _____

   (b) Incandescent light _____

2. (a) Hydrogen source (H):

V	B	A	G	Y	O	R

   R = red, O = orange, Y = yellow, G = green, A = aqua, B = blue, V = violet

   (Draw approximate position of line and write the scale reading on line.)

   (b) Mercury source (Hg):

V	B	A	G	Y	O	R

   (c) Sodium Source (Na):

V	B	A	G	Y	O	R

   (d) What type of spectra is given off by these sources? _____

3. Percentage error for:

   (a) H_α _____ (red line)

   (b) H_β _____ (aqua line)

   (c) Na _____ (yellow doublet)

4. Types of spectra produced by fluorescent light source:

   (a) _____

   (b) _____

   (c) Explain why these are present: _____

       _____

       _____

   (d) What type of gas is found in fluorescent fixtures? _____

5. Answer to questions in item 5 of Procedure: _____

   _____

   _____

   _____

EXPERIMENT 28

RELATIVE HUMIDITY

Materials Needed

Wet- and dry-bulb hygrometer
Calorimeter cup
Ice
Thermometer
Hair hygrometer
Alcohol (rubbing)

Background

Some moisture is present in the atmosphere at all times. Body functions, cooking, and evaporation from bodies of water are some of the ways water vapor is added to the atmosphere. Water condensing on a cold glass, the formation of dew and fog at ground level, and cloud formation show evidence of this moisture. Muggy days make us very conscious of this moisture. Air is said to be *saturated* when it contains all the vapor it can hold. The temperature at which this occurs is called the *dew point*. At higher temperatures more water vapor is required for saturation. At dew point, if the temperature is lowered or more vapor is added, condensation will occur.

   Relative humidity is the ratio of the amount of water vapor in a unit volume to the amount contained in the same volume at saturation. The terms of this ratio may be expressed in grams of water vapor per unit volume. The same value will result if the ratio is expressed as the ratio of the pressure of the vapor at this time to the pressure the vapor would exert at the same temperature if it were saturated. With a relative humidity of less than 3 percent, a feeling of dryness may result. A value of 80 percent gives what is called a *muggy* feeling. Relative humidity is usually expressed in percent. Our bodily comfort depends a great deal on the relative humidity.

   Several instruments called *hygrometers* have been developed to determine the relative humidity. A common method is to take readings from wet- and dry-bulb thermometers. Using these readings and prepared tables, the relative humidity can be determined. The most common instrument for home use is called a *hair hygrometer*.

There are seven quantities involved in various hygrometric determinations:

d = absolute humidity, the number of grams of water vapor in a cubic meter of space

D = density at saturation at the same temperature

p = pressure exerted by the water vapor in the air at the density d

P = pressure that would be exerted by the water vapor if the air were saturated at the same temperature

t_d = dew point, the temperature at which saturation will take place if the amount of water vapor in the air remains unchanged

t = temperature of the air

r = relative humidity

Two of the fundamental relations of r are:

$$r = \frac{d}{D}$$

or, since the pressure the water vapor exerts is proportional to the density,

$$r = \frac{p}{P}$$

Procedure

To find the dew point, select a small calorimeter cup with a polished surface. Fill the cup about one-third full of water. Insert a thermometer. Add small pieces of ice until a film begins to form on the outside of the cup. Watch the film with care and read the thermometer to tenths of a degree. Have the cup in a place where there are no air currents and do not breathe on the cup. Keep the water stirred.

Table 28-1 Vapor pressure of water

Temperature, °C	Pressure, mmHg	Temperature, °C	Pressure, mmHg
0	4.6	55	118.0
5	6.5	60	149.2
10	9.2	65	187.5
15	12.7	70	233.8
20	17.4	75	289.1
25	23.6	80	355.5
30	31.5	85	433.6
35	42.2	90	526.0
40	55.0	95	633.9
45	71.9	100	760.0
50	92.2		

Table 28-2 Absolute humidity table

Temperature, °C	Humidity, g/m³	Temperature, °C	Humidity g/m³	Temperature °C	Humidity, g/m³
0	4.84	12	10.59	24	21.54
1	5.18	13	11.25	25	22.80
2	5.54	14	11.96	26	24.11
3	5.92	15	12.71	27	25.49
4	6.33	16	13.50	28	26.93
5	6.76	17	14.34	29	28.45
6	7.22	18	15.22	30	30.05
7	7.70	19	16.14	31	31.70
8	8.21	20	17.12	32	33.45
9	8.76	21	18.14	33	35.27
10	9.33	22	19.22	34	37.18
11	9.93	23	20.35	35	39.18

Remove the ice and stir lightly until the film begins to disappear. Repeat several times until you are able to bring the temperature of appearance to within 1° of the temperature of disappearance. Finally, average all results and record. Record the room temperature. Determine the relative humidity from the dew point and a table of temperatures and corresponding pressures of saturated water vapor and absolute humidity.

Tables 28-1 through 28-3 give a list of tempertures from 0°C to 100°C with the pressure of saturated water vapor in millimeters of mercury (mm Hg) for each temperature and the density in grams per cubic meter (g/m³) for saturated water vapor. Having determined the dew point, the relative humidity is readily computed from such tables. The method is illustrated with an example. Suppose the measured dew point is 5°C. If the atmosphere were saturated, from Table 28-1 the vapor pressure would be 6.5 mmHg. Although the air is not saturated, the same vapor pressure is exerted. Assume the room temperature to be 20°C. If the atmosphere were saturated, the pressure would be 17.4 mmHg. The ratio of the pressures would be 6.5/17.5 or 0.37. The relative humidity then would be 37 percent. If the value of absolute humidity (shown in Table 28-2) were used, the value of r would be 6.76/17.12 or 39%.

To find the relative humidity from a wet- and dry-bulb hygrometer, two thermometers are used. The bulb of one is covered with an absorbent cloth that dips into a dish of water. The instrument should be used in moving air. This condition can be realized by fanning the instrument for a minute or two (some instruments, such as the sling psychrometer, are made to rotate about an axle). The readings of both the wet and dry thermometers are recorded. Tables are prepared for computing the relative humidity directly from these readings. See Table 28-3.

Readings of the dry-bulb thermometer (room temperature) are given in the left column. The difference between the wet- and dry-bulb readings is given across the top. The intersection of the two lines gives the relative humidity in percent. That is, the dry bulb reading were 20°C (room temperature) and the wet bulb reading 15°C, the difference would be 5°C. Using the dry-bulb reading of 20°C and going across to the difference column marked 5°C, the relative humidity is 58 percent.

Find the relative humidity using the wet- and dry-bulb readings. Compare the results from the two methods. Check with a hair hygrometer.

A high relative humidity reduces the rate of evaporation, giving a smaller difference between the readings of the two thermometers. To observe the cooling effect due to evaporation, place a small amount of alcohol on the back of your hand. As the alcohol evaporates, you

Table 28-3 Relative humidity (in percent)

Dry bulb, °C	Difference of Dry-Bulb Minus Wet-Bulb Temperature, °C															
	.5	1	1.5	2	2.5	3	3.5	4	4.5	5	7.5	10	12.5	15	17.5	20
−15	79	79	58	38	18											
−12.5	82	65	47	30	13											
−10	85	69	54	39	24	10										
−7.5	87	73	60	48	35	22	10									
−5	88	77	66	54	43	32	21	11								
−2.5	90	80	70	60	50	42	37	22	12							
0	91	82	73	65	56	47	39	31	23	15						
2.5	92	84	76	68	61	53	46	38	31	24						
5	93	86	78	71	65	58	51	45	38	32						
7.5	93	87	80	74	68	62	56	50	44	38						
10	94	88	82	76	71	65	60	54	49	44	19					
12.5	94	89	84	78	73	68	63	58	53	48	25					
15	95	90	85	80	75	70	66	61	57	52	31	12				
17.5	95	90	86	81	77	72	68	64	60	55	36	18				
20	95	91	87	82	78	74	70	66	62	58	40	24				
22.5	96	92	87	83	80	76	72	68	64	61	44	28	14			
25	96	92	88	84	81	77	73	70	66	63	47	32	19			
27.5	96	92	89	85	82	78	75	71	68	65	50	36	23	12		
30	96	93	89	86	82	79	76	73	70	67	52	39	27	16		
32.5	97	93	90	86	83	80	77	74	71	68	54	42	30	20	11	
35	97	93	90	87	84	81	78	75	72	69	56	44	33	23	14	
37.5	97	94	91	87	85	82	79	76	73	70	58	46	36	26	18	10
40	97	94	91	88	85	82	79	77	74	72	59	48	38	29	21	13

will note a cooling effect. Blow across your hand. You will note a greater cooling effect. This is due to more rapid evaporation. Heat is required to evaporate a liquid; this heat comes from your body, making you feel cooler. The more rapid the evaporation, the greater the cooling effect. One readily notes this cooling effect after getting out of a swimming pool, especially on a windy day.

EXPERIMENT 28

RELATIVE HUMIDITY
REPORT SHEET

Name _____   Section _____

Temperature of room   _____

Average temperature at which film forms   _____

Average temperature at which film disappears   _____

Dew point   _____

Absolute humidity at room temperature (D)   _____

Absolute humidity at dew point (d)   _____

Relative humidity $r = d/D$ _____

Reading of dry-bulb thermometer   _____

Reading of wet-bulb thermometer   _____

Relative humidity   _____

Reading of hair hygrometer   _____

EXPERIMENT 29

INTRODUCTION TO WEATHER MAPS

Materials Needed

Weather maps (included)
Weather map explanatory sheet (student or school to obtain)*

*Available from Climate Analysis Center, Room 808, World Weather Building, Washington, DC 20233.

Background

All of you are acquainted with weather maps since you see them on television or in the paper each day. Weather maps are prepared by the National Weather Service and are used in weather forecasting. Such maps are plotted daily with the aid of a computer. Weather data are collected from all over the United States and sent to a central location. Stations are indicated on the weather map. The data indicating the type of weather are plotted around a circle. This definite arrangement of the weather data is known as the *station model*. A complete explanation of the symbols and codes used in plotting weather maps is published in an Explanation of the Weather Map. A great deal of information can be displayed in this manner.

Procedure

The station models on the weather maps are rather small. Two station models, shown in the figure below, have been enlarged to assist you in interpreting the models.

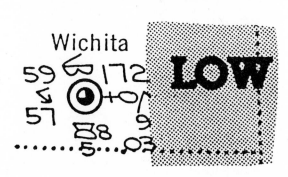
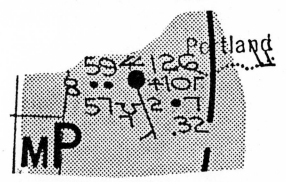

Refer to the explanation sheet of the weather map for the symbols and order of arrangement used in plotting the weather on the station model. From these data determine the weather for two stations.

EXPERIMENT 29

INTRODUCTION TO WEATHER MAPS
REPORT SHEET

Name _____   Section _____

	Wichita	Portland
1. Station name, iii		
2. Amount of sky covered by clouds, N		
3. Direction from which wind is blowing, dd		
4. Wind velocity, ff		
5. Visiblity, VV		
6. Present Weather, ww		
7. Past weather, W		
8. Barometric pressure, PPP		
9. Air temperature, TT		
10. Sky covered by low or middle clouds, N_h		
11. Cloud type (low), C_L		
12. Height of cloud base, h		
13. Cloud type (middle), C_M		
14. Cloud type (high), C_H		
15. Dew point, $T_d T_d$		
16. Barometric tendency, a		
17. Pressure change in 3 h, pp		
18. Amont of precipitation, RR		
19. Time precipitation began or ended, R_t		
20. Depth of snow on ground S		

Problems

Referring to the Portland Weather Station, what does the MP mean? _____

Are you surprised at the cloud coverage and the heavy clouds? _____

Would you expect rain? _____

Weather maps for August 2, 3, and 4, 1970, appear on pages 264–266. They show the path of a very destructive hurricane that hit the Texas coast in that year.

1. See the August 2 weather map.

   (a)   What is the type of front just north of the Great Lakes? _____

   (b)   What is happening to it? (See the August 3 and 4 maps.) _____

   (c)   Note the low in central Canada. (August 2 map, north of Trout Lake). What is the lowest pressure at its center? _____

   (d)   Note the high just north of Montana on the August 2 map. What is the maximum pressure? _____

   (e)   What would you expect about the pressure, temperature, and humidity at this high? _____

   (f)   Is this verified at Edmondton? _____

2. Record the pressure at Corpus Christi for each date:

   (a)   August 2 _____

   (b)   August 3 _____

   (c)   August 4 _____

   (d)   How do you account for these changes? _____

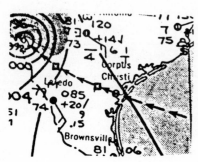
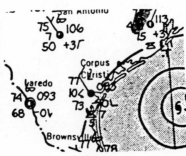
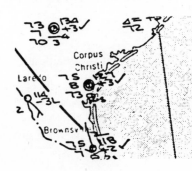

3. What kept Celia moving in a western direction? See the August 2 map. (Note the pressures east and west of Celia.)

4. How do you account for the weather along the front just south of the Great Lakes on August 3?

5. What is the kind of front off the northwest coast on August 4?

6. Find the speed of the High in south-central Canada on August 2 and 3. (Use the scale in the lower left-hand corners of the maps.)

7. How fast did Celia move on August 3? Use the dotted line to find an average speed.

8. Generally there is a great deal of rain near a hurricane area. How do you account for the entire south being dry on August 2?

9. Find the latitude and longitude of Celia on Monday, August 3.

10. The weather map gives the weather for what time of day?

11. What is the air pressure in the southeast corner of Montana on August 3?

12. Which way is the air rotating in the center of Celia?

13. On August 3 in San Francisco is the weather wet or dry?

14. Note the scarcity of weather stations in Mexico. Is this a help or hinderance to U.S. weather forecasters?

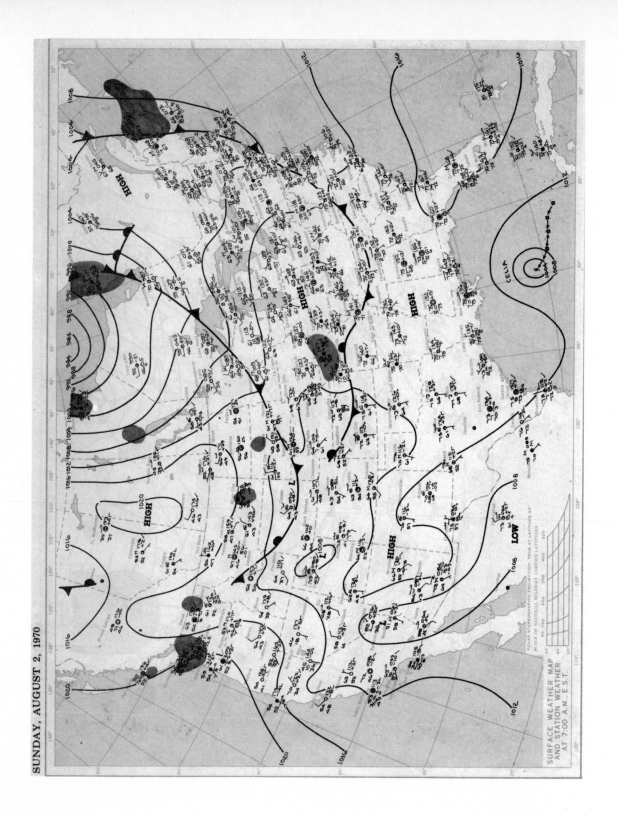

SUNDAY, AUGUST 2, 1970

SURFACE WEATHER MAP
AND STATION WEATHER
AT 7:00 A.M. E.S.T.

MONDAY, AUGUST 3, 1970

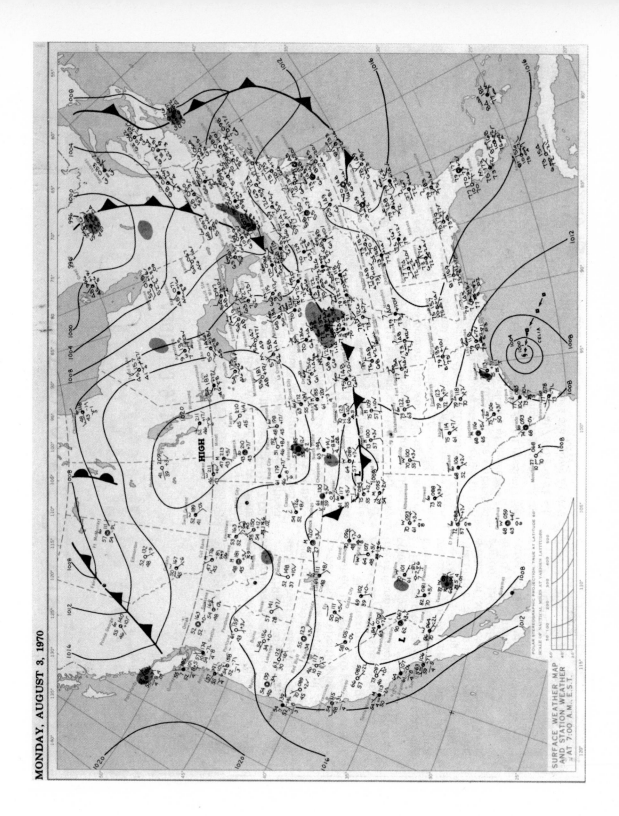

SURFACE WEATHER MAP
AND STATION WEATHER
AT 7:00 A.M. E.S.T.

265

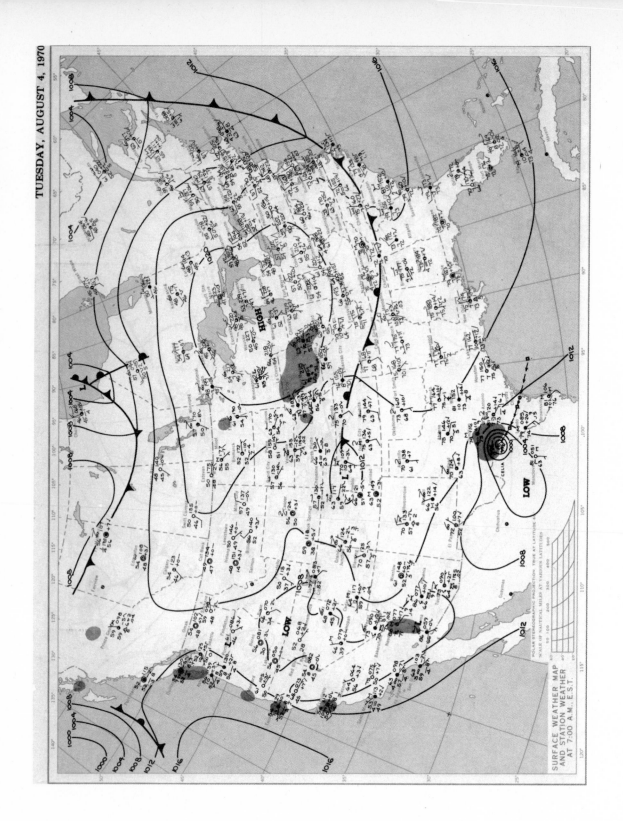

SURFACE WEATHER MAP
AND STATION WEATHER
AT 7:00 A.M., E.S.T.

The Coded Message

Plot the following coded weather messages around a station circle. Use the symbolic model and the sample plotted report on the Explanation of the Weather Map.

Code:

1.   405   61825   16638   08838   67378   38736   72580